ARITHMETIC:

A Review

J. LOUIS NANNEY

RICHARD D. SHAFFER

Mathematics Department

Miami-Dade Junior College

JOHN WILEY & SONS, INC. NEW YORK · LONDON · SYDNEY · TORONTO

10 9 8 7 6

Library of Congress Catalog Card Number: 75–93297

SBN 471 62990 1
Printed in the United States of America.

PREFACE

Some people, either for lack of opportunity or lack of interest, are weak in the fundamentals of arithmetic. Others have once felt secure in these fundamentals, but time has erased this feeling. This text has been written for the specific purpose of teaching the basic facts of arithmetic, which are approached from a structural or algebraic point of view. All skills necessary at the pre-algebra level are covered in this text. An Introduction to Algebra is also included.

This is a "write-in" text. Space is provided for the working of all problems, and the student is urged to work every one. Furthermore, this book has been written to be read. Every word is important. Blanks left in many of the explanations are to be filled in by the student.

When the student finishes this text, every word should have been read, every blank filled in, and every problem worked. Only then will full benefit have been received.

Miami, Florida
June 1969

J. Louis Nanney
Richard D. Shaffer

CONTENTS

Chapter 1

READING AND WRITING COUNTING NUMBERS

Numbers represented by 0, 1, 2, 3, 4, and so on, are called the Counting Numbers. This is a very important set of numbers. In a study of mathematics, all other sets of numbers can be defined in terms of the counting numbers.

The student should become proficient in all operations of the counting numbers before attempting to study other sets of numbers.

What counting number should be used to indicate the following?

a, a, a, a	________	a's in this set?
a, a, a	________	a's in this set?
b, b, b, b, b	________	b's in this set?
	________	c's in this set?

Do you see why 0 is considered a counting number?

Strictly speaking, "number" is an abstract idea and "numeral" is any symbol used to represent a number.

To write a numeral for any Counting Number we use only the digits 0, 1, 2, 3, 4, 5, 6, 7, 8, 9. Counting numbers larger than nine are written by the use of two or more digits.

For the number three hundred sixty four, we write the numeral 364.

$$\begin{aligned} 364 &= 3 \text{ hundreds} + 6 \text{ tens} + 4 \text{ ones} \\ &= 300 + 60 + 4 \end{aligned}$$

Notice that the digits have a positional value or place value.

From right to left the three positional values are ones, tens, hundreds.

hundreds, tens, ones

All counting numbers are written so that this group of ones, tens, and hundreds is repeated. Each group of three numerals has a name. We will use H, T, and O to stand for Hundreds, Tens, and Ones.

hundreds tens ones *hundreds tens ones* *hundreds tens ones*

H T O	H T O	H T O	H T O
billion	million	thousand	

Notice that the right-most group is not named. This group has a name, the *units* group, but we do not name it when reading a numeral.

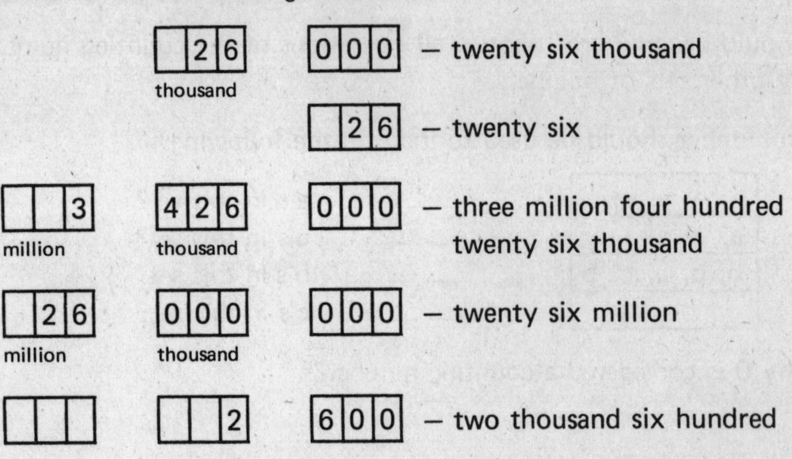

EXERCISES

Write the names of the following numerals.

1. 203 ______________________________
2. 203 000 ______________________________
3. 1 472 ______________________________
4. 12 001 ______________________________
5. 30 060 ______________________________
6. 186 441 ______________________________
7. 100 100 ______________________________
8. 4 902 000 ______________________________
9. 1 500 000 000 ______________________________
10. 37 403 612 019 ______________________________
11. 2 846 370 000 ______________________________
12. 21 900 376 004 ______________________________

Every group of three digits—ones, tens, hundreds—has a name. It is not likely that you will need to know the name for any group above billions.

EXERCISES

Write the name for the given numerals or the numerals for the given names.

1. 5,280 ______________________________

____

2. 52,800 ______________________________

____

3. 7,001 ______________________________

____

4. 40,300 ______________________________

____

5. 134,001 ______________________________

____

6. 123,456,134 ______________________________

____

7. ______________________ Five thousand forty

8. ______________________ One hundred thirty seven thousand eighty four

9. ______________________ Four hundred thirty one million

10. ______________________ Three billion four hundred million seven hundred thirteen thousand

11. ______________________ Eight hundred million eight hundred thousand eight hundred eight

12. ______________________ Sixty three thousand four

13. ______________________ Seven billion nine hundred forty three thousand five hundred thirty two

Chapter 2

ADDITION AND SUBTRACTION OF COUNTING NUMBERS

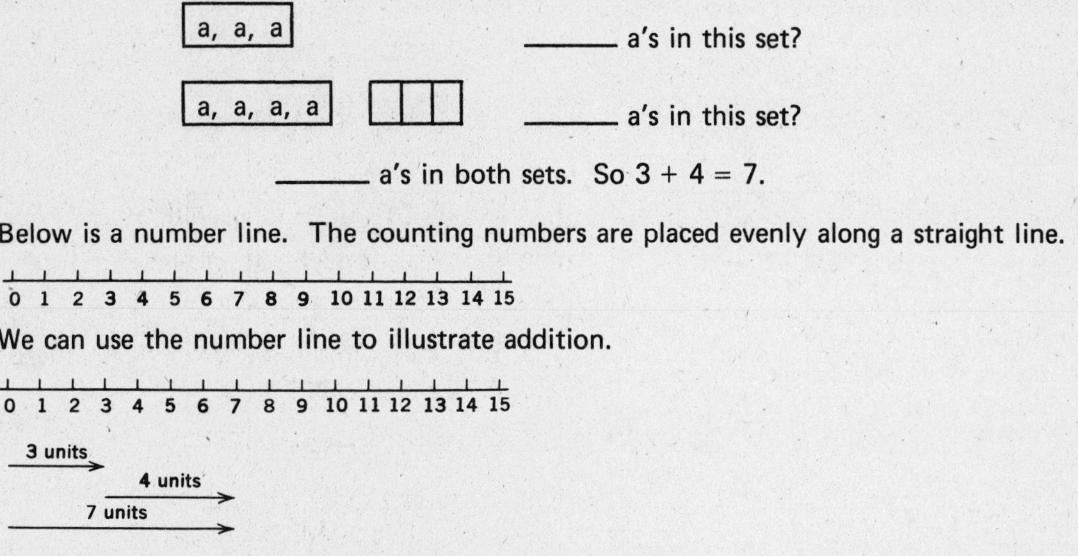

______ a's in this set?

______ a's in this set?

______ a's in both sets. So 3 + 4 = 7.

Below is a number line. The counting numbers are placed evenly along a straight line.

We can use the number line to illustrate addition.

These simple illustrations show us the meaning of addition of counting numbers. The operation of addition gives us the *sum* or *total* of two counting numbers.

Facts we know

1. There are only ten digits.
2. All counting numbers are written by using one or more of these digits.
3. All numerals are written by using a group of three digits having place values of ones, tens, hundreds.
4. Ten ones = one ten
   Ten tens = one hundred

Conclusion

To add any two counting numbers it is only necessary that we know the sum of all possible combinations of the ten digits, taken 2 at a time.

Example: If we know that $3 + 4 = 7$
Then we know that $30 + 40 = 70$
$300 + 400 = 700$
$13 + 4 = 17$
$23 + 4 = 27$
$803 + 24 = 827$

Because it is necessary that we know only all possible combinations of the ten digits, we should know them thoroughly. They should be learned so well that the combinations are automatic, so that we never have to stop and think.

Complete the following addition table.

+	0	1	2	3	4	5	6	7	8	9
0										
1										
2										
3										
4										
5										
6										
7										
8										
9										

SOME FACTS ABOUT THE ADDITION TABLE

There are ten rows, each of which contains ten sums. How many sums in all?

____

There is one numeral that has no effect when it is added to any other numeral. That numeral is ____________ .

Several patterns are evident from the table. For instance, reading the main diagonal from upper left to lower right is actually counting by two's. Find and list some other patterns.

____

____

____

____

____

____

____

____

Did you find other patterns? One of the most important ones you could have found is that the table is *balanced around* (or symmetric with respect to) the main diagonal from upper left to lower right. For instance, if we folded the table along a line from the upper left corner to the lower right corner the numbers falling on each other would be alike.

This balance or symmetry is due to facts such as $5 + 3 = 8$ and $3 + 5 = 8$.

This property of addition is called the *Commutative Property of Addition* and is stated in precise terms as

$$A + B = B + A$$

when A and B are counting numbers.

In words, this property states that when we add numbers the order of addition will not change the sum.

A CLOSE LOOK AT THE ADDITION ALGORITHM
(Algorithm Means "Method")

What really happens when we add two counting numbers?

Let's work an addition problem together.

Problem: Add 574 and 299.

574 = 5 hundreds + 7 tens + 4 ones
299 = 2 hundreds + 9 tens + 9 ones
7 hundreds + _ tens + _ ones

Can we now write the answer as 7 hundreds + 16 tens + 13 ones? ____________

We must remember that:

10 tens = 1 hundred and
10 ones = 1 ten

So, 16 tens = 10 tens + 6 tens
= 1 hundred + 6 tens

and 13 ones = 10 ones + 3 ones
= 1 ten + 3 ones

We then have

7 hundreds + 16 tens + 13 ones =
7 hundreds + 1 hundred + 6 tens + 1 ten + 3 ones =
8 hundreds + 7 tens + 3 ones = 873

So, 574 + 299 = 873

Now let's look at the same problem in a shorter, more common method.

  574 First we add 9 + 4 and get 13. We recognize this as 1 ten and 3 ones.
+ 299

So we record the 3 ones and "carry" the one ten to the tens place.

$$\begin{array}{r} 1 \\ 574 \\ +\ \underline{299} \\ 3 \end{array}$$

Now we add 9 + 7 + 1 and get 17. This is 17 tens or 1 hundred + 7 tens.

We record the 7 tens in the tens place and "carry" the 1 hundred to the hundreds place.

$$\begin{array}{r} 11 \\ 574 \\ +\ \underline{299} \\ 73 \end{array}$$

We now add 2 + 5 + 1 and get 8. This is 8 hundreds.

We record the 8 in the hundreds place and the problem is complete.

$$\begin{array}{r} 11 \\ 574 \\ +\ \underline{299} \\ 873 \end{array}$$

The amount "carried" does not need to be recorded if you can remember.

EXERCISES

Find the following sums.

1. $\begin{array}{r} 34 \\ \underline{99} \end{array}$

2. $\begin{array}{r} 40904 \\ \underline{89498} \end{array}$

3. $\begin{array}{r} 417 \\ 320 \\ \underline{933} \end{array}$

4. $\begin{array}{r} 987 \\ \underline{86} \end{array}$

5. $\begin{array}{r} 1239 \\ \underline{427} \end{array}$

6. $\begin{array}{r} 994 \\ 397 \\ \underline{116} \end{array}$

7. $\begin{array}{r} 9099 \\ \underline{9199} \end{array}$

8. $\begin{array}{r} 8034 \\ \underline{8034} \end{array}$

9. $\begin{array}{r} 1342 \\ 306 \\ \underline{297} \end{array}$

10. $\begin{array}{r} 30692 \\ \underline{9874} \end{array}$

11. $\begin{array}{r} 7362 \\ \underline{988} \end{array}$

12. $\begin{array}{r} 4482 \\ 3693 \\ \underline{14975} \end{array}$

Test yourself

This is a test designed to be worked in six minutes. Answers are upside down at the bottom of the page. Time and check yourself.

1. 367
284
26
913
88

2. 5002
3764
196
916
24

3. 97846
23900
4162
96840

4. 97
83
61
24
65
51
63
98
12

5. 384
961
255
803
612
983

6. 421
578
111
235
612
263

7. 407
260
3712
4089
616
2340

8. 96
208
41
723
91
106
84

9. 472
906
41
963
811
181
406

10. 284
631
96
4963

11. 48 + 216 + 15 + 12 = ____________

12. 2314 + 987 + 65 = ____________

Answers: 1. 1678 2. 9902 3. 222748 4. 554 5. 3998 6. 2220 7. 11424 8. 1349 9. 3780 10. 5974 11. 291 12. 3366

SUBTRACTION OF COUNTING NUMBERS

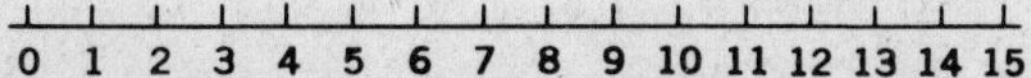

This is a number line. The counting numbers are placed evenly along a straight line. We will use the number line to illustrate subtraction.

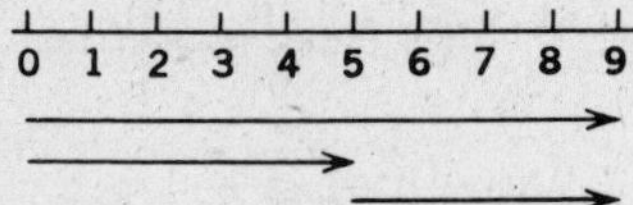

The above line shows that $9 - 5 = 4$.

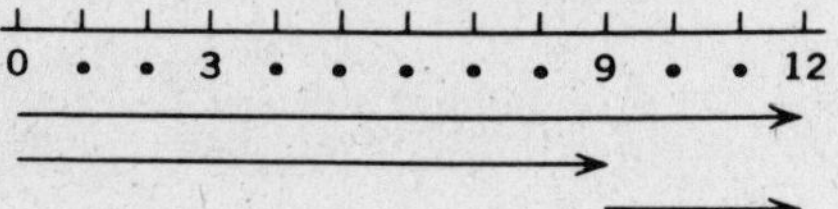

This line shows that $12 - 9 = 3$. Count the spaces from 9 to 12. There are ________ spaces.

These illustrations show us that subtraction of counting numbers means "finding the difference." If A, B, and C are counting numbers, $A - B = C$ means that C is the "distance" from B to A. Or, to be more precise, C is the number that must be added to B to get A.

EXERCISES

Draw arrows to indicate the following subtractions.

1. $12 - 3$

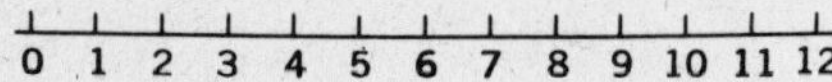

2. $15 - 14$

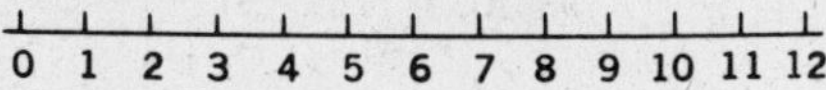

3. $9 - 9$

0 1 2 3 4 5 6 7 8 9 10 11 12

Facts we know

1. All counting numbers are written by using the ten digits 0, 1, 2, 3, 4, 5, 6, 7, 8, 9.
2. If we know all "differences" of these digits we can subtract.

Do we now need a subtraction table? Do we need to memorize all possible differences of the digits?

To answer the question about the need for a subtraction table, let's look more closely at a simple problem.

$9 - 5 = 4$. The difference of 9 and 5 is 4. But we already know that $5 + 4 = 9$.

How are these facts related? Is it true that if we know all addition facts, we also know the subtraction facts? ________

Example: $12 - 7 = 5$ because $7 + 5 = 12$

$15 - 7 = \_$ because $7 + \_ = 15$

$18 - 8 = \_$ because $8 + \_ = 18$

$36 - 9 = \_$ because $9 + \_ = 36$

$a - b = c$ because $b + \_ = \_$

Is $9 + \_\_ = 16$ a subtraction problem or an addition problem? ____________

It should be clear that addition facts are sufficient (all that is needed) for subtraction.

EXERCISES

Find the missing number.

1. $12 - \_\_ = 4$
2. $7 + \_\_ = 14$
3. $26 - 9 = \_\_$
4. $7 + \square = 15$

   $\square = \_\_$
5. $18 - 9 = \_\_$

   $\_\_ = \_\_$
6. $x + 6 = 14$

   $x = \_\_$
7. $7 - x = 1$

   $x = \_\_$

Most of the problems in the preceding exercise were such that the answer could come directly from an addition table.

Let's now look at some more difficult subtraction problems.

Problem: Subtract 412 from 576.

576 = 5 hundreds + 7 tens + 6 ones
412 = 4 hundreds + 1 tens + 2 ones

Subtracting, we get 1 hundred + 6 tens + 4 ones = 164.

So: 576 − 412 = ________ .

Problem: Subtract 383 from 724.

724 = 7 hundreds + 2 tens + 4 ones
383 = 3 hundreds + 8 tens + 3 ones

We subtract 3 ones from 4 ones and get 1 one. But we cannot subtract 8 tens from 2 tens. It now becomes necessary to rewrite 724. Remembering that 1 hundred = 10 tens, we can write:

724 = 6 hundreds + 12 tens + 4 ones.

Now we can subtract 383.

724 = 6 hundreds + 12 tens + 4 ones
383 = 3 hundreds + 8 tens + 3 ones

3 hundreds + 4 tens + 1 one

So: 724 − 383 = ________ .

This process of rewriting a number for subtraction is sometimes called "borrowing."

Let's look at the same problem using a more common method of writing it:

Problem: Subtract 383 from 724.

724
383

First we subtract 3 ones from 4 ones and record this difference in the ones place.

$$\begin{array}{r} 724 \\ \underline{381} \\ 1 \end{array}$$

Now we notice that 8 tens cannot be subtracted from 2 tens so we borrow 1 hundred from the 7 hundreds. We can strike through the 7 and write a 6 above it to show that we have borrowed. Since we borrowed 1 hundred = 10 tens, we now have 12 tens that we can record in the tens column.

$$\begin{array}{r} 6\ 12\ \ \\ \not{7}\ \not{2}\ 4 \\ \underline{3\ 8\ 3} \\ 1 \end{array}$$

Now we can finish the subtraction problem.

$$\begin{array}{r} 612 \\ \not{7}\not{2}4 \\ \underline{383} \\ 341 \end{array}$$

EXERCISES

Subtract.

1. $\begin{array}{r} 824 \\ \underline{711} \end{array}$
2. $\begin{array}{r} 1674 \\ \underline{1593} \end{array}$
3. $\begin{array}{r} 271 \\ \underline{199} \end{array}$
4. $\begin{array}{r} 27{,}863 \\ \underline{208} \end{array}$
5. $\begin{array}{r} 4164 \\ \underline{2286} \end{array}$
6. $\begin{array}{r} 921 \\ \underline{700} \end{array}$
7. $\begin{array}{r} 635 \\ \underline{277} \end{array}$
8. $\begin{array}{r} 9876 \\ \underline{6789} \end{array}$
9. $\begin{array}{r} 884 \\ \underline{848} \end{array}$

Remember from the statement on page 11 that A − B = C is the same as A = C + B.

This provides us with a "check" for subtraction. In each exercise on page 14 add your answer to the subtrahend (number subtracted) and see if you get the minuend (number from which you subtracted).

Check

1.	2.	3.	4.	5.
------	--------	------	---------	--------
711	1593	199	208	2286
824	1674	271	27863	4164

6.	7.	8.	9.
------	------	--------	------
700	277	6789	848
921	635	9876	884

Subtraction problems in which zeroes appear in the minuend must be given special attention or mistakes may result.

Example: Subtract 388 from 507.

507
388

Since 8 ones cannot be subtracted from 7 ones, we need to borrow a ten. (1 ten = 10 ones) But we have 0 tens. This means we must first rewrite 507.

507 = 5 hundreds + 0 tens + 7 ones
= 4 hundreds + 10 tens + 7 ones
= 4 hundreds + 9 tens + 17 ones

The problem looks like this when we show our borrowing.

```
 4 9 17
 5 0 7      17 − 8 = 9;  9 − 8 = 1;  4 − 3 = 1
 3 8 8
 -----
 1 1 9
```

So: 507 − 388 = ________

Check: 119 + 388 = 507

Always make a record of your borrowing, because if a mistake is made, you need to be able to retrace your steps in computation to determine the type of error you have made.

EXERCISES

Subtract.

1. 850 − 721
2. 580 − 278
3. 3002 − 564
4. 634 − 127
5. 134 − 79
6. 396 − 396
7. 14,206 − 3,147
8. 14,206 − 3,104
9. 2007 − 304
10. 2007 − 398
11. 1427 − 1372
12. 612 − 94
13. 9046 − 389
14. 173,421 − 79,963
15. 49,608 − 2,139
16. 1634 − 467
17. 430,006 − 18,994
18. 2,046,192 − 324,186
19. 13,789,420 − 7,149,390
20. 123,456,789 − 69,947,791

Chapter 3

NUMBER SENTENCES

Check True or False for each of these.

	T	F
1. 6 + 4 = 10	____	____
2. 15 – 2 = 13	____	____
3. 8 + 7 = 15	____	____
4. 2 + 6 = 7	____	____
5. 3 – 1 = 1	____	____
6. 8 + 3 = 11	____	____
7. 4 + 6 =	____	____

What did you check for problem 7? If you left it blank, you have recognized an essential (necessary) quality of a *mathematical sentence.*

A mathematical sentence must be either true or false.

Which of these are sentences?

1. $8 + 6 = 14$
2. $7 + 5 = 3$
3. $9 + 6 = 15$
4. 9 is a counting number
5. $4 + 8$
6. $3 - (7 - 5)$
7. $8 + (4 + 2) = 14$
8. the number 723

Let us now examine a special type of mathematical expression.

$$5 + x = 8$$

Is this a sentence? Your answer should be "no" since it is neither true nor false as it now stands. With a replacement for the place holder "x" this becomes a sentence.

If $x = 4$ the sentence is $5 + 4 = 8$ and is false. Nevertheless it is a sentence.

If $x = 3$ the sentence is $5 + 3 = 8$ and is true.

The fact that $5 + x = 8$ is neither true nor false is characteristic of what we refer to as "open sentences." Whether it is true or false is "open" and depends on the replacement for the place holder.

Which of the following are open sentences?

1. $4 + 7$
2. $4 + 7 = x$
3. $3 + 8 = 11$
4. $9 - x = 4$
5. $16 + x$

Exercises 1 and 5 are number "expressions" but are not sentences or open sentences because they are neither true nor false, nor can a replacement of the placeholder make them either true or false. Exercise 3 is a sentence as it stands.

Our concern in open sentences is always the same. It is to *find the replacements for the placeholder* that will make the open sentence a *true* sentence.

In the following exercise find replacements for the place holders that will make each open sentence true.

1. $5 + x = 9$

   $x =$ ____

2. $x + 6 = 15$

   $x =$ ____

3. $N - 5 = 12$

   $N =$ ____

4. $216 + 512 = N$

   $N =$ ____

5. $164 - N = 45$

   $N =$ ____

6. $164 + 45 = x$

   $x =$ ____

7. $5004 - N = 3125$

   $N =$ ________

8. $x - 6 = 62$

   $x =$ ____

9. $17 + x = 23$

   $x =$ ____

10. $14 + 8 = x$

    $x =$ ____

11. $217 - 14 = x$

    $x =$ ____

12. $44 + 22 + x = 68$

    $x =$ ____

13. $37 + 3 + x = 112$

    $x =$ ____

14. $5040 - 3178 = N$

    $N =$ ________

REVIEW TEST OF CHAPTERS 1 – 4

1. Write names for the following numerals.

   (a) 7,916 ____________________

   (b) 90,090 ____________________

2. What is the sum of fourteen thousand three; nine thousand four hundred four; twenty one; and seven hundred seven?

3. Find the sums.

   (a)
   ```
    5614
    2806
    6183
     204
    5193
   ```

   (b)
   ```
   76004
    2016
    3908
    4300
     207
   ```

4. Find the differences.

   (a)
   ```
     98614
   –  9948
   ```

   (b)
   ```
     7000
   –  484
   ```

   (c)
   ```
     9642
   – 7882
   ```

5. Find the replacement that will make each open sentence true.

   (a) $19 + x = 41$

   x = ____

   (b) $86 + x + 23 = 162$

   x = ____

   (c) $98 - x = 74$

   x = ____

   (d) $1762 - x = 83$

   x = ____

PUNCTUATION IN NUMBER SENTENCES

Parentheses ( ), brackets [ ], and braces { }, are used in number sentences to group numbers together. When numbers are placed in parentheses they are to be considered as a single number. For instance, $5 + (6 - 2)$ would indicate that some number is added to 5. Or $7 - (3 + 2)$ would indicate that some number is subtracted from 7.

Rule: If parentheses, brackets, or braces are used in a number expression or sentence, operations on the numbers enclosed must be done first.

Example: $17 - (3 + 4) = x$

$17 - 7 = x$

$\_\_\_\_ = x$

Example: $(17 - 3) + 4 = x$

$14 + 4 = x$

$\_\_\_\_ = x$

These two examples clearly indicate that placement of parentheses can make a difference in an answer.

EXERCISES

Solve each of the following problems. Then rewrite the problem by moving the parentheses to give a different answer.

1. $(12 - 4) + 2 = x$ ____________________

   $x = \_\_\_\_$

2. $(18 - 6) - 3 = x$ ____________________

   $x = \_\_\_\_$

3. $(15 - 3) - 3 = x$ ____________________

   $x = \_\_\_\_$

Place parentheses in the following problems to make them *true* number sentences.

1. $8 - 3 + 5 = 0$
2. $8 - 3 + 5 = 10$
3. $15 + 6 - 5 = 16$
4. $141 - 8 + 3 = 130$
5. $141 - 8 - (3 + 2) = 138$
6. $141 - 8 - 6 - 2 = 137$
7. $12 - 8 - 3 + 5 = 2$
8. $12 - 8 - 3 + 5 = 12$
9. $32 + 7 - 4 = 35$
10. $32 - 7 - 4 = 29$

In the previous exercises you may have needed to use a double set of parentheses in the same problem. More than one set of parentheses in a problem can be confusing and can be avoided by using brackets [ ], or braces { }. For example, in exercise 5, 141 – 8 – (3 + 2) = 138 we know that 141 – 3 = 138. We also know that the given parentheses indicate that (3 + 2) must be given a value first. The problem is then 141 – 8 – 5 = 138. It is clear that 141 – (8 – 5) = 138 is the desired form, so the correct answer is 141 – (8 – (3 + 2)) = 138. Since both sets of parentheses make the writing awkward, we should write 141 – [8 – (3 + 2)] = 138. This form leaves no doubt as to the operations desired.

When two or more sets of parentheses (or brackets or braces) are used in a number sentence or expression we must evaluate the *innermost* set first.

Example: 5 + {17 – [8 – (3 –2)]} = x

5 + {17 – [8 – 1]} = x

5 + {17 – ____ } = x

5 + ____ = x

x = ____

One agreement is now necessary: if more than two numbers appear in a number expression involving only addition and subtraction and *no* parentheses are used, the operations are to be performed as they appear in order from *left* to *right.*

Example: 17 + 6 – 2 + 3 = x

23 – 2 + 3 = x

21 + 3 = x

x = ____

Use of parenthesis is more desirable, but if this rule is followed we can be assured of getting the intended answer.

EXERCISES

In each of the following problems find x:

(a) by performing the operations as they appear from left to right

(b) by placing parentheses around the first two numbers

(c) by placing parentheses around the last two numbers

1. $x = 12 - 6 + 3$ (a) x = ____ (b) x = ____ (c) x = ____

2. $x = 18 + 12 - 4$ (a) x = ____ (b) x = ____ (c) x = ____

3. $41 - 31 - 7 = x$ (a) x = ____ (b) x = ____ (c) x = ____

4. $56 + 7 + 2 = x$ (a) x = ____ (b) x = ____ (c) x = ____

5. $28 - 18 - 10 = x$ (a) x = ____ (b) x = ____ (c) x = ____

Find x in each of the following:

6. $x = 4 + 12 - (6 + 2)$ ____________

7. $x = 86 - 18 - 6 + (5 - 4)$ ____________

8. $x = 100 - 50 + 81 - (21 + 10)$ ____________

9. $20 - 3 + (80 + 7) = x$ ____________

Now that we have parentheses with which to work, we need to discuss another very important property of addition. This property is called the *Associative Property of Addition* and is stated in symbols as A + (B + C) = (A + B) + C.

In words this property states that if three numbers are to be added, the sum will be the same regardless of the grouping of the numbers.

Place parentheses in this sentence to illustrate the associative property:

5 + (4 + 3) = 5 + 4 + 3

5 + ______ = ___ + ___

______ = ______

Notice that the terms are different but the sums are the same.

Some problems in the previous set of exercises were instances of the Associative Property of Addition. Notice that in problem 4 the answers for parts (b) and (c) were the same. (56 + 7) + 2 = 56 + (7 + 2) = 65. However, the answers for (b) and (c) in problem 5 were not the same. This should clearly emphasise that the associative property is a property of *addition,* not of the numbers themselves.

EXERCISES

Place parentheses in the following to group them so that they are easy for you to add mentally, then write the sum.

1. 27 + 3 + 16 =

   ____ + ____ = ______

2. 184 + 7 + 3 =

   ____ + ____ = ______

3. 17 + 8 + 5 =

   ____ + ____ = ______

4. 420 + 80 + 17 =

   _____ + _____ = ______

The two special properties of addition that we have studied are the commutative property and the associative property.

A + B = B + A	commutative property
(A + B) + C = A + (B + C)	associative property

EXERCISES

In the blanks at the right, write the word *associative* or *commutative* to explain which property has been used.

1. (3 + 4) + 5 = 5 + (3 + 4) ____________________
2. 3 + (4 + 5) = (3 + 4) + 5 ____________________
3. [(5 + 7) + 2] + 3 = (5 + 7) + (2 + 3) ____________________
4. [(5 + 7) + 2] + 3 = 3 + [(5 + 7)] + 2 ____________________

Many times we use both the associative and commutative properties in one step of a computation. When we rearrange *and* regroup in one step, we are using the commutative-associative principle.

Use this principle in the following to find the sums as easily as possible.

5. 17 + 7 + 3 = (17 + ___) + ___ = ______
6. 14 + 18 + 6 + 2 = ____________________ = ______
7. 9862 + 4000 + 8 = ____________________ = ______
8. 1463 + 294 + 7 + 6 = ____________________ = ______
9. 1600 + 210 + 400 = ____________________ = ______
10. 7 + 2 + 1 + 83 = ____________________ = ______

CHANGING WORD PROBLEMS TO NUMBER SENTENCES

Let's look at the number sentence $x + 5 = 8$. Can this same number sentence be written in another way? Of course $8 - 5 = x$ is the same sentence since x is the same number in both cases.

The number sentence $x + 5 = 8$ can be written many different ways as one or more English sentences. For instance: John has five dollars. How much more does he need to buy an eight-dollar shirt?

You would probably think about this problem in this manner: amount John has + unknown amount = amount needed, or $5 + x = 8$.

You could also think of it this way: amount needed − amount John has = unknown amount, or $8 - 5 = x$.

The important thing to remember is that *the number sentence must state the same facts that are stated by the English sentence.*

Actually we could say that we are *translating* from one language to another. We are translating from *English sentences* to *number sentences.*

Most difficulties that arise in verbal problems come from improper translation of English words and phrases to numbers or symbols.

EXERCISES

Write a number phrase equivalent to each English phrase.

1. Eight less than twenty ____________________
2. Four hundred sixteen diminished by seven ____________________
3. The sum of nine and seven ____________________
4. The difference of eight and three increased by two

   ____
5. Forty six diminished by the sum of two and seven

   ____

EXERCISES

Translate each phrase or sentence to a number expression.

1. 5 more than x ____________________
2. 17 increased by 200 ____________________
3. The difference of 15 and x ____________________
4. The sum of 9 and y ____________________
5. x increased by 16 ____________________
6. 2 less than x ____________________
7. y equals x increased by 16 ____________________
8. A is 20 more than B ____________________
9. Subtract 5 from the sum of 2 and 7 ____________________
10. The sum of 8 and 2 is the same as the difference of x and A.

    ____

11. Johnny's score of 80 was ten less than Bill's score of x.

    ____

12. x is 14 diminished by 3 ____________________
13. The sum of 2 and 12 increased by x ____________________
14. The sum of 5 and the difference of 7 and 2

    ____

Let's work this word problem together: Mr. Smith bought a used car priced at $900. The sales tax was $27, the tag cost $22, and the registration fee was $2. The dealer allowed Mr. Smith $250 for his old car. How much money did Mr. Smith need to make the trade?

After carefully reading the problem we first decide on a basic fact that simplifies our task.

Total cost less the trade-in equals money needed.

Total cost: Price of car + tax + tag + registration fee or

900 + ___ + ___ + ___

Trade-in: _____

So our number equation becomes:

900 + 27 + 22 + 2 − 250 = x

_____ − 250 = x

_____ = x

Mr. Smith needs $ _____ .

Example: John had 85 cents. He paid 35 cents for his school lunch, 15 cents for his cold drink, and 10 cents for a bar of candy. He found a nickel on the sidewalk. How much more does he need for a 50-cent movie ticket?

Price of ticket − money on hand = money needed.

Money needed = x
Price of ticket = 50 cents
money on hand = original amount + amount found − amount spent
= 85 + 5 − (____ + ____ + ____)

So x = 50 − [(85 + 5) − (______)]

x = _____

The "punctuation" (parentheses and brackets) is necessary to properly write this problem as a number sentence.

Make a number sentence for each of the following verbal problems; then find the answer by solving the number sentence.

1. Mr. Brown had a balance of $87 in his bank account. He deposited $110 more and then wrote checks for $12, $8, $27, and $54. What is his new balance?

2. The Browns are going on a trip of 864 miles. If they drive 221 miles the first day and 386 miles the second day, how far are they from their destination?

3. How much more than the difference of 18 and 12 is the sum of 18 and 12?

4. The Allegheny and Ohio rivers together are 1, 306 miles long. If the Allegheny is 325 miles long, how long is the Ohio?

5.

Ocean	Area in sq mi	Average depth in ft
Pacific	63,801,668	14,048
Atlantic	31,839,306	12,880
Indian	28,356,276	13,002

From this table answer the following questions.

(a) How much deeper is the Pacific than the Atlantic, on an average?

____

(b) How much larger is the Pacific in square miles than the Indian?

____

(c) Is the Pacific larger than the Atlantic and Indian together? If so, how much larger?

____

____

6. Tokyo has the largest population of any city in the world. In 1960 the population of Tokyo was 11,021,579. In 1960 New York City had a population of 7,781,984. How many more people lived in Tokyo than in New York City?

____

7. George Washington was born in 1732 and became President in 1789. How old was he when he became President?

____

8. The moon is 221,463 miles from the earth at perigee (closest approach to earth) and is 252,710 miles away at apogee (farthest distance). How much farther away is the moon at apogee than at perigee?

____

9. The population of the United States in 1800 was 5,308,483; in 1850 it was 23,191,876; in 1900 it was 75,994,574; and in 1950 it was 150,697,361.

   (a) What was the population increase from 1800 to 1900?

   ____

   (b) What was the population increase from 1850 to 1950?

   ____

   (c) Which 100 years had the greatest increase?

   ____

   (d) If the increase from 1950 to 2000 is the same as from 1800 to 1950, what will be the population in the year 2000?

   ____

EXERCISES

1. The perimeter of a plane geometric figure is the distance around it. Find the perimeter of the following figures.

(a)

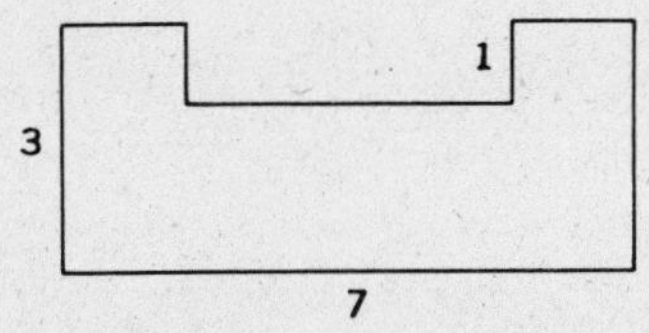

P = ______________

(b)

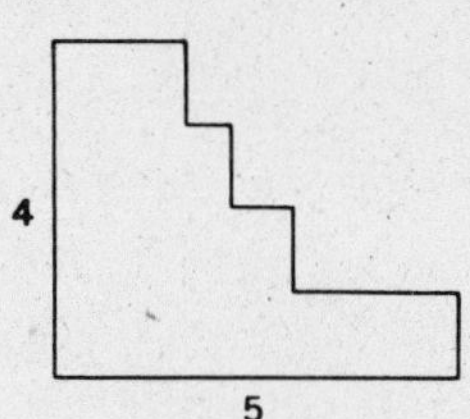

P = ______________

2. If A and B are counting numbers, fill the blanks in this table.

	A	B	A + B	A − B
(a)	8	7	____	____
(b)	9	____	15	____
(c)	24	____	____	13
(d)	17	____	20	____
(e)	14	____	23	____
(f)	____	8	____	4
(g)	____	6	24	____
(h)	____	3	____	18
(i)	8	____	8	____

3. Mr. Iley and his neighbor wish to buy a plane. The type they want has a list price of \$6750.00. They have found a used plane they can buy for \$3800.00, but it needs some repairs. Adding a radio would cost \$650; a new magneto would cost \$157; two new tires would cost \$30 each; a flashing beacon would cost \$83; and repainting would cost \$64. If Mr. Iley can do all labor himself, how much will they save by buying the used plane?

____

4. The following patterns of numbers can be continued by using addition or subtraction. Fill the blanks.

   (a) 2, 5, 8, 11, ____, ____, ____

   (b) 1, 4, 9, 16, ____, ____, ____

   (c) 1, 1, 2, 3, 5, 8, ____, ____, ____

   (d) 2, 7, 6, 11, 10, 15, 14, ____, ____, ____

   (e) 1, 6, 5, 11, 9, 16, 13, ____, ____, ____

5. The following array of numbers can be continued if you discover the pattern. It is based on addition. Fill the blanks.

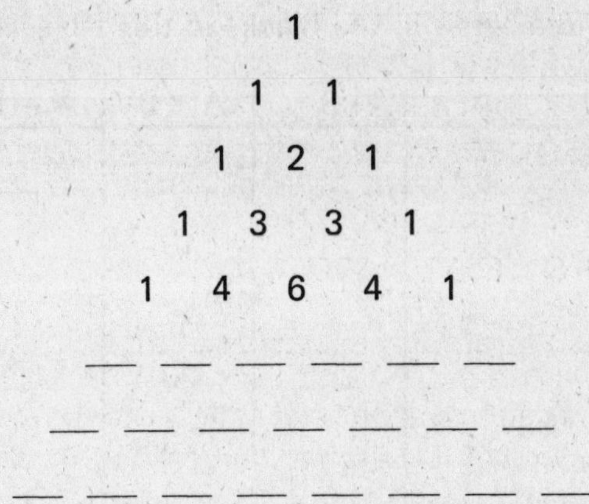

6. Find the perimeter of this figure if all segments are equal.

Chapter 4

MULTIPLICATION AND DIVISION OF COUNTING NUMBERS

In previous chapters we have dealt only with sums and differences of numbers. We now wish to introduce another operation – multiplication. The operation of multiplication will be indicated in one of three ways: using the symbol "x" (for example, 7×8); using the symbol " · " (for example, $7 \cdot 8$); or using the convention of parentheses [for example, (7) (8)], that is, parentheses written together with no operation sign between them will always indicate multiplication. Each of these examples is read "seven times eight."

When we state that $7 \times 8 = 56$ we mean that eight used as an addend seven times is fifty six. That is $8 + 8 + 8 + 8 + 8 + 8 + 8 = 56$.

44×23 means 23 used as an addend ____ times.

The operation of multiplication on whole numbers is not necessary. But think of the trouble we would have if we wished to multiply 564 by 843. It would be necessary to use 843 as an addend ____ times. So, even though multiplication is not necessary it is a very desirable operation to master.

Let's recall a basic fact from Chapter 1: "Every counting numeral may be written by using the ten digits 0 through 9." With this in mind we see that mastering the operation of multiplication depends on our knowledge of the product of any two of the ten digits.

Complete the following multiplication table.

	0	1	2	3	4	5	6	7	8	9
0										
1										
2				6						
3										
4										
5			10							
6								42		
7					28					
8										
9										

It cannot be emphasized too strongly that you *must* know the facts of this table if you are to be proficient in multiplication. You should know these facts so thoroughly that they come to you automactically, without you having to stop and think. So study the table, repeating it over and over until this is true.

EXERCISES

1. Multiplication by 0 always gives _____ .
2. Multiplication by _____ doesn't change the number being multiplied.
3. Is the table symmetric to (i.e., balanced around) either diagonal? _______

   Why? ____

   ____

   ____

Your answer to Exercise 3 should be that the table is balanced around the diagonal from upper left to lower right because the order of multiplication does not change the answer (product). For example, $7 \times 8 = 8 \times 7$.

This property of multiplication is called the *commutative* property and is written in symbols as *$A \times B = B \times A$ for all numbers A and B.*

Another important property of multiplication is illustrated by these examples:

$(2 \times 4) \times 5 = \_\_\_\_ \times 5 = \_\_\_\_\_$.

$2 \times (4 \times 5) = 2 \times \_\_\_\_ = \_\_\_\_\_$.

This is called the *associative* property of multiplication and is written in symbols as *$(A \times B) \times C = A \times (B \times C)$ for all numbers A, B and C.*

Notice that the commutative property involves a change in order, (commute means to move from one place to another) whereas the associative property is only a change in the way the numbers are associated or grouped.

If A, B, and C are numbers and $A \times B = C$, C is called the "product" of A and B.

You will recognize that these two properties of multiplication have the same names as two properties we discussed for addition.

EXERCISES

Fill these blanks.

1. associative property of addition ____________________
2. associative property of multiplication ____________________
3. commutative property of addition ____________________
4. commutative property of multiplication ____________________

(Notice the likeness and differences of 1 and 2; also of 3 and 4.) In exercises 5 – 7 fill the blank with the property used.

5. $(3 + 4) \times 7 = 7 \times (3 + 4)$ ____________________
6. $(7 \times 6) \times 8 = 7 \times (6 \times 8)$ ____________________
7. $[(3 + 2) \times 7] = (3 + 2) \times (7 \times 2)$ ____________________

Each of the four properties in the previous exercises involves only one operation at a time. There is another property that we now wish to discuss, which is very important and yet is not quite so obvious as the four already discussed. This property involves both multiplication and addition and is named the *distributive* property. In symbols it is stated as $A \times (B + C) = A \times B + A \times C$.

Some examples of the distributive property are:

1. $5 \times (2 + 6) = 5 \times 2 + 5 \times 6$

   $5 \times 8 = 10 + 30$

   $40 = 40$

2. $7 \times (8 + 2) = 7 \times 8 + 7 \times 2$

   $7 \times 10 = 56 + 14$

   $70 = 70$

3. $25 \times (6 + 4) = 25 \times 6 + 25 \times 4$

   $25 \times 10 = \_\_\_\_ + \_\_\_\_$

   $250 = 250$

If you look closely at these examples you will recognize that the computation may be easier on one side than on the other. This property can many times be used to make computation simpler.

EXERCISES

Compute these as they are grouped and decide which way you think is easier in each case. Do the computation mentally if possible.

1. Find 22×23 by:
   (a) $22 \times (20 + 3)$ and (b) $(22 \times 20) + (22 \times 3)$

2. Find 33×12 by:
   (a) $33 \times (10 + 2)$ and (b) $(33 \times 10) + (33 \times 2)$

3. (a) $112 \times (6 + 4)$ (b) $(112 \times 6) + (112 \times 4)$

4. (a) $3 \times (14 + 6)$ (b) $(3 \times 14) + (3 \times 6)$

EXERCISES

(a) Compute and (b) tell what property is illustrated.

	(a)	(b)
1. $(5 + 8) + 7 = 5 + (8 + 7)$	______	______________________
2. $7 \times (2 + 18) = 7 \times 2 + 7 \times 18$	______	______________________
3. $14 \times (3 + 7) = (3 + 7) \times 14$	______	______________________
4. $12 + 8 = 8 + 12$	______	______________________
5. $(13 + 7) + 14 = 14 + (13 + 7)$	______	______________________
6. $12\,(20 + 5) = 12 \times 20 + 12 \times 5$	______	______________________

THE MULTIPLICATION ALGORITHM

Let's take a close look at a simple multiplication problem.

$$24 \times 38$$

This could be given as 38 used as an addend ______ times. But 38 can also be written as 30 + ______ . So the problem can be interpreted as 30 used as an addend ______ times *plus* 8 used as an addend ______ times. This is the basis for our usual method of multiplying numbers of more than one digit. You should recognize this as the distributive property applied twice.

$$24 \times 38 = 24 \times (30 + 8)$$

$$= (24 \times 30) + (24 \times 8)$$

$$\text{But } (24 \times 30) = (20 + 4) \times 30 = (20 \times 30) + (4 \times 30)$$

$$\text{and } (24 \times 8) = (20 + 4) \times 8 = (\_\_ \times \_\_) + (\_\_ \times \_\_)$$

$$\text{So } 24 \times 38 = (20 \times 30) + (4 \times 30) + (20 \times 8) + (4 \times 8)$$

$$= 600 + 120 + 160 + 32$$

$$= \_\_\_\_\_\_\_\_$$

Now let's look at the familar way of multiplying and see how it is an application of the distributive property.

```
     24
  x  38
  -----
     32  :  we multiply 8 x 4
    160  :  we multiply 8 x 20
    120  :  we multiply 30 x 4
    600  :  we multiply 30 x 20
  -----
    912  :  we add the products.
```

Compare this with the illustration at the bottom of the previous page.

We can make the foregoing multiplication shorter by doing some steps mentally. The different steps are numbered and explained.

```
Step 1.   24
          38
         ---
           2
```

1. We multiply 8 x 4 and get 32, but only write the 2, remembering that we have 30 remaining.

```
Step 2.   24
          38
         ---
         192
```

2. We multiply 8 x 20 and get 160, but we had 30 remaining, which we add to give 190. With the 2 already written this makes 192.

```
Step 3.   24
          38
         ---
         192
          20
```

3. We multiply 30 x 4 and get 120, write the 20, and remember that we have 100 remaining.

```
Step 4.   24
          38
         ---
         192
         720
```

4. We multiply 30 x 20 and get 600, but we had 100 remaining, and with the 20 already written this gives 720.

```
Step 5.   24
          38
         ---
         192
         720
         ---
         912
```

5. We add the partial products to get the total product of 912.

Sometimes the 0 is left out in the second partial product like this:

```
   24
   38
  ---
  192
  72
  ---
  912
```

This is permissible, since the sum of 0 and a number leaves the number unchanged. *But* if zero is left out, a space must be left for it.

If you make a note of the digit being carried your work will look like this:

```
  1
  3
 24
 38
---
192
72
---
912
```

EXERCISES

Find the following products.

```
1.   87        2.   23        3.  821        4.  1314
   x 92           x 57          x  26          x   28

5.  212        6.  823        7.  591        8.  9988
  x 161          x 712          x 223          x  976

9.  846       10.  946       11.  7862      12.   961
  x  50          x  51          x  111          x 876
```

When the numbers being multiplied contain zeros, special care must be taken to avoid mistakes.

Example

```
  361
x 203
 1083
```

For the first partial product we multiply 3 x 361

```
  361
x 203
 1083
 000
```

Multiplying by zero gives only zero.

```
  361
x 203
 1083
 000
722
73283
```

We now multiply by 2 and add the partial products to get our answer.

The row of zeros is not necessary since it does not change the sum of the partial products. *However, the zero does hold a position.*

```
  361
x 203
 1083
7220
73283
```

To avoid writing a row of zeros, simply place one zero to hold the place value.

Example

```
   3002
 x 2004
  12008
600400
6016008
```

= 6,016,008

EXERCISES

1. $\begin{array}{r} 586 \\ \times\ 203 \\ \hline \end{array}$

2. $\begin{array}{r} 7704 \\ \times\ 809 \\ \hline \end{array}$

3. $\begin{array}{r} 8030 \\ \times\ 500 \\ \hline \end{array}$

4. $\begin{array}{r} 2695 \\ \times\ 3003 \\ \hline \end{array}$

5. $\begin{array}{r} 9876 \\ \times\ 1011 \\ \hline \end{array}$

6. $\begin{array}{r} 216 \\ \times\ 30 \\ \hline \end{array}$

7. $\begin{array}{r} 908 \\ \times\ 908 \\ \hline \end{array}$

8. $\begin{array}{r} 7654 \\ \times\ 909 \\ \hline \end{array}$

9. $\begin{array}{r} 7272 \\ \times\ 6060 \\ \hline \end{array}$

10. $\begin{array}{r} 365 \\ \times\ 96 \\ \hline \end{array}$

11. $\begin{array}{r} 2840 \\ \times\ 612 \\ \hline \end{array}$

12. $\begin{array}{r} 9283 \\ \times\ 707 \\ \hline \end{array}$

13. $N = 17 \times 54$

    $N =$

14. $X = 3 \times (51 \times 20)$

    $X =$

15. A farmer stored 72 bales of hay that averaged 163 pounds for each bale. How many pounds of hay did he store?

16. 186 people are on a ship and the average weight per person is 153 pounds. There is also a cargo of 412 cartons of produce weighing 63 pounds per carton. If the load limit of the ship is 60,000 pounds, is the load over or under the limit, and by how much?

17. The volume of a rectangular solid, such as a box, is found by multiplying the length, width, and height. Volume is measured in cubic units. If "V" is the volume, "L" is length, "W" is width, and "H" is height, then the formula for volume is V = LWH.

(a) Find the volume of this rectangular solid.

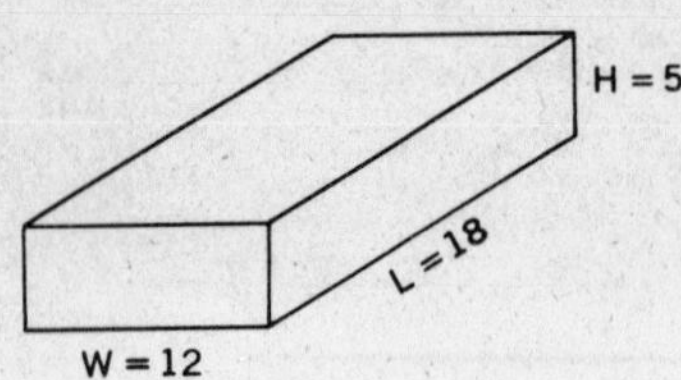

V = L x W x H

V = ____ x ____ x ____

V = ______ cubic inches

(b) What is the volume of a rectangular solid if L = 5 ft, W = 3 ft, H = 3 ft?

(c) What is the volume of a solid if L = W = H = 8 ft? (Such a figure is called a cube.)

DIVISION OF COUNTING NUMBERS

Division of counting numbers is commonly indicated in any one of three ways.

$772 \div 112$

$112\overline{)772}$

$\dfrac{772}{112}$

Regardless of which way the problem is written, it is read as "seven hundred seventy two divided by one hundred twelve."

$772 \div 112$ is a way of asking "How many sets of 112 are contained in 772?" We will first look at a simple division problem to discover the meaning of the operation of division.

$20 \div 5$ (How many sets of 5 in a set of 20?)

To answer this question we will subtract 5 from 20 as many times as possible and count the number of subtractions.

$20 - 5 =$ ____; $15 - 5 =$ ____; $10 - 5 =$ ____; $5 - 5 =$ ____ .

We found it possible to subtract 5 from 20 how many times? This means there are ____ sets of 5 in a set of 20, or $20 \div 5 = 4$.

Division can thus be defined as *repeated subtraction.*

EXERCISES

By repeated subtraction, solve the following.

1. $772 \div 112 =$ ______

   $772 - 112 =$ ______; $660 - 112 =$ ______

   ____ $- 112 =$ ______; . . .

2. $1083 \div 215 =$ ______ with a remainder of ______ .

   $1083 - 215 =$ ______; . . .

Let us now look for shortcuts to do this repeated subtraction.

$$112 \overline{)784}$$

How many sets of _____ are in a set of _____?

We could find the answer by subtracting 112 from 784 until there are no more sets of 112 left. We will do the repeated subtraction but we will subtract several sets of 112 at the same time. To do this let's guess an answer. Suppose we decide to guess that there are 5 sets of 112 in 784.

If we multiply 112 by 5 we get _____. So 5 sets of 112 are in 560.

784 – 560 = 224; so 784 contains 5 sets of 112 plus 224.

$$\begin{array}{r} 5 \\ 112 \overline{)784} \\ \underline{560} \\ 224 \end{array}$$

We now need to find how many sets of 112 are in 224. We guess an answer of 2.

2 x 112 = 224; so 224 contains 2 sets of 112.

This means that 784 contains 5 sets of 112 plus 2 sets of 112 or 7 sets of 112.

$$\begin{array}{r} 7 \\ \overline{2} \\ 5 \\ 122 \overline{)784} \\ \underline{560} \\ 224 \\ \underline{224} \\ 0 \end{array} \qquad 784 \div 122 = 7$$

In working such problems we attempt to guess as close as possible to the answer so that the amount of work will be reduced. It never really matters if our guess is too small, but if it is too large we must make a new one. Our guess is correctly called a "trial divisor."

EXERCISES

Do the following division problems.

1. $26\overline{)2158}$

2. $9568 \div 92$

3. $\dfrac{2058}{21}$

4. $1458 \div 54$

5. $27\overline{)50625}$

6. $81\overline{)2187}$

7. $64\overline{)2752}$

8. $1188 \div 22$

9. $11,144 \div 14$

10. $33\overline{)2904}$

In the previous set of problems there were no remainders. This is not always true. Let us now do another example step by step and see what happens when there is a remainder.

Problem: $9863 \div 84$

```
84 | 9863
```

Our first guess is that 9863 contains at least 100 sets of 84.

```
      100
84 | 9863
     8400
     ----
     1463
```

We multiply 100 × 84 and subtract 8400 from 9863.

Our new problem is $1463 \div 84$. But we do not show this as a different problem, but continue our guessing in the same frame work.

```
       10
      100
84 | 9863
```

Our new guess is 10. We multiply 10 by 84 and subtract 840 from 1463.

```
        5
       10
      100
84 | 9863
     8400
     ----
     1463
      840
     ----
      623
      420
     ----
      203
```

Our guess now is 5. We multiply 5 by 84 and subtract 420 from 623.

```
        2
        5
       10
      100
84 | 9863
     8400
     ----
     1463
      840
     ----
      623
      420
     ----
      203
      168
     ----
       35
```

It is obvious that our guess was too small, since 203 is greater than 84. We guess again. Our guess now is 2, so we multiply 2 × 84 and subtract 168 from 203.

We have now determined that there are 100 + 10 + 5 + 2 = 117 sets of 84 contained in a set of 9863. We have also determined that there is a set of 35 left over. This is called the remainder.

$$9863 = 84 \times 117 + 35$$

It is not necessary to place the zeros in our trial divisors. However, it is necessary that the numbers be kept in their proper position.

```
      117    R. 35
84 ) 9863
     84
     ---
     146
      84
     ---
      623
      588
      ---
       35
```

This is a shorter way to write the problem, but if the zeros are an aid, the student should not hesitate to use them.

EXERCISES

Work each of these problems. Write each answer in the form of the example (i.e., 9863 = 117 x 84 + 35).

1. 84) 98675

2. 306) 11423

3. 1861) 96483

4. 762) 148721

5. 27) 270273

6. 921) 100345

We have described multiplication as repeated addition, and division as repeated subtraction. This indicates that the same kind of relation between these two operations should exist between addition and subtraction, and this is true. We now state this relationship in symbols:

$$\text{if } \frac{A}{B} = C \text{ then } A = B \times C$$

Example: Since $\frac{56}{8} = 7$ then $56 = 7 \times 8$

We can also state that since $7 \times 8 = 56$ then $\frac{56}{8} = 7$ and $\frac{56}{7} = 8$.

Or in symbols: if $A \times B = C$ then $A = \frac{C}{B}$ and $B = \frac{C}{A}$.

This last statement has one exception.

What is the value of $\frac{14}{0}$? If $\frac{14}{0} = A$ then $14 = A \times 0$ — but $A \times 0 = 0$ for all values of A.

So we see there is no value for $\frac{14}{0}$. In fact in all of mathematics, *division by 0 is not defined.*

This relationship between division and multiplication provides us with an excellent method of checking division.

Example: Check to see if $9828 \div 84 = 117$.

If $\frac{9828}{84} = 117$ then it should be true that $9828 = 117 \times 84$.

Check:

$$\begin{array}{r} 117 \\ \times\ 84 \\ \hline 468 \\ 936 \\ \hline 9828 \end{array}$$

Example: Does $9863 \div 84 = 117$ with remainder 35?

If $\frac{9863}{84} = 117$ with remainder 35 then $9863 = (84 \times 117) + 35$.

Exercises

Divide and check.

1. $13\overline{)544569}$

2. $108\overline{)756216}$

Chapter 5

FACTORS AND PRIMES

A divisor, or factor, of a counting number is any counting number that evenly divides the given number (i.e., leaves a remainder of 0).

Two is a factor or divisor of 12, since 12 ÷ 2 leaves a remainder of 0.

Other divisors of 12 are 1, 2, 3, ___, ___, ___.

Divisors of 36 are 1, 2, 3, 4, ___, ___, ___, ___ and 36.

There is one counting number that can never be used as a divisor. It is ___.

Is 1 a divisor of every counting number? ___

If "A" is a counting number, what *two* counting numbers are divisors of A? ___ and ___. Your answer should be A and 1 since every nonzero counting number can be divided by itself as well as the 1.

Remember that zero is an exception. It can *never* be used as a divisor.

List some divisors of the following numbers.

1. 20: 1, 2, 4, ___, ___, ___
2. 30: ___, ___, ___, ___, ___, ___, ___, ___
3. 6: ___, ___, ___, ___
4. 4: ___, ___, ___
5. 5: ___, ___
6. 7: ___, ___
7. 21: ___, ___, ___, ___
8. 37: ___, ___
9. 38: ___, ___, ___, ___
10. 10: ___, ___, ___, ___
11. 17: ___, ___
12. 0: ________________
13. 35: ________________
14. 3: ________________
15. 2: ________________
16. 14: ________________
17. 200: ________________
18. 8: ________________
19. 1: ________________
20. 121: ________________

You have noticed from these problems that the given number is always a divisor of itself. That is, in problem 1 you listed 20 as a divisor of 20; in problem 2 you listed 30 as a divisor of 30. What other number have you listed as a divisor of each number? ________. Zero is again an exception, but notice that zero is divisible by any number except itself.

You may also notice that some of the numbers in the problems had only two divisors. These were 5, 7, 37, 17, ____, and ____.

These are special numbers and are called *primes.* There are many other numbers of this type but in all cases *the prime numbers are those counting numbers with exactly two different divisors.*

The following is a list of a few of the prime numbers.

Supply the missing primes:

2	29
3	____
____	37
____	____
____	43
13	47
17	____
____	59
23	61

The following numbers are *not* prime. Refer to the underlined description of a prime number above, then state why these are not prime.

Answers

1. 6 This number has more than 2 divisors. They are: 1, 2, 3, 6
2. 24 ____
3. 4 ____
4. 0 ____
5. 21 ____
6. 33 ____
7. 1 ____
8. 8 ____
9. ½ ____
10. 121 ____

The prime numbers will prove so useful to us that we should consider a method for finding these numbers.

A famous mechanical device for locating primes is the "Sieve of Eratosthenes." We arrange the counting numbers in an array for convenience and eliminate those that do not have exactly two different divisors and are therefore left with only primes. The way this device works is outlined here. We will find the primes less than 24.

~~1~~	2	3	~~4~~
5	~~6~~	7	~~8~~
~~9~~	~~10~~	11	~~12~~
13	~~14~~	~~15~~	~~16~~
17	~~18~~	19	~~20~~
~~21~~	~~22~~	23	~~24~~

One can be marked out first since it has only one divisor and is not prime.

The next numeral 2 is not marked out but now, all of the other even numbers, that is 4, 6, 8, 10, 12, 14, 16, 18, 20, 22, and 24, may be crossed out since each will have 3 or more divisors. For example, the divisors of 4 are 1, 2 and 4. It is not prime.

Three is not crossed out but the numbers 6, 9, 12, 15, 18, 21, and 24 are marked out since each of these has 3 as a divisor. (We will later call these numbers multiples of 3.) Skip numbers that have already been crossed out and go to the next number that is not marked out.

Five is the next number not marked out, so we mark out 10, 15, and 20, which have 5 as a divisor, meaning that they will have *more* than two divisors and are therefore *not* prime. These are to be marked out, but in this case we find them to be eliminated already since they have previously been found to have more than two divisors.

We need not look further for nonprime numbers in this set. Five, when multiplied by itself is 25, a numeral larger than 24, which is the largest number in the table.

In summary: to find all prime numbers less than a counting number N, we (1) list the numbers from 1 to N (and immediately strike out 1) then (2) each time we come to a number not eliminated we strike out all numbers divisible by it. (3) The process ends when a number is not eliminated which when multiplied by itself is equal to or greater than N. (4) All numbers not thus eliminated will be prime.

You will be asked to make a sieve of Eratosthenes to find all primes less than 200. Since 15 × 15 is greater than 200, the process will end when you reach 15 or a number not eliminated which is greater than 15.

Should 1 be crossed out? _____

Use the method described on the previous page to locate all primes in this table, *then* list these prime numbers on the right hand margin of this page for future use. You should find 46 prime numbers.

										Primes	
1	2	3	4	5	6	7	8	9	10		
11	12	13	14	15	16	17	18	19	20		
21	22	23	24	25	26	27	28	29	30		
31	32	33	34	35	36	37	38	39	40		
41	42	43	44	45	46	47	48	49	50		
51	52	53	54	55	56	57	58	59	60		
61	62	63	64	65	66	67	68	69	70		
71	72	73	74	75	76	77	78	79	80		
81	82	83	84	85	86	87	88	89	90		
91	92	93	94	95	96	97	98	99	100		
101	102	103	104	105	106	107	108	109	110		
111	112	113	114	115	116	117	118	119	120		
121	122	123	124	125	126	127	128	129	130		
131	132	133	134	135	136	137	138	139	140		
141	142	143	144	145	146	147	148	149	150		
151	152	153	154	155	156	157	158	159	160		
161	162	163	164	165	166	167	168	169	170		
171	172	173	174	175	176	177	178	179	180		
181	182	183	184	185	186	187	188	189	190		
191	192	193	194	195	196	197	198	199	200		

COMPLETE FACTORIZATION

A number is factored by changing it to a *product* of two or more other numbers. Some numbers may be written as products in several ways. For example:

(a) $8 = 8 \cdot 1$

or

(b) $8 = 2 \cdot 4$

or

(c) $8 = 2 \cdot 2 \cdot 2 \cdot 1$

or

(d) $8 = 4 \cdot 2$

The factorization in (d) is the same as (b) because of the commutative property for multiplication, so we will call these factorizations the same.

Factorizations of 48 are:

$48 = 6 \cdot 8$
$48 = 3 \cdot 4 \cdot 4$
$48 = 1 \cdot 48$
$48 = 2 \cdot 2 \cdot 2 \cdot 2 \cdot 3$
48 = ______________
48 = ______________ } write other factorizations here
48 = ______________

Factorization is important in both arithmetic and algebra. We use this process to simplify fractions, to add or subtract fractions, to solve equations, and to study numbers.

Have you noticed that each of the factors of a number is a divisor of the number? Remember however, that the factors are *multiplied* together and equal the given number.

Show one way in which these numbers can be factored.

1.	$18 = \underline{2 \cdot 9}$	11.	$2 =$ _______
2.	$27 =$ _______	12.	$63 =$ _______
3.	$4 =$ _______	13.	$64 =$ _______
4.	$5 =$ _______	14.	$65 =$ _______
5.	$30 =$ _______	15.	$1 =$ _______
6.	$100 =$ _______	16.	$121 =$ _______
7.	$12 =$ _______	17.	$42 =$ _______
8.	$37 =$ _______	18.	$225 =$ _______
9.	$10 =$ _______	19.	$4a =$ _______
10.	$49 =$ _______	20.	$1024 =$ _______

The process of writing a number as a *product* of *prime* factors is called *prime factorization.* That is, each of the factors is a prime number.

The prime factorization of 4 is:

$$4 = 2 \cdot 2$$

Since prime numbers are to be used here, list again the *first 10 prime* numbers.

____ ____ ____ ____ ____ ____ ____ ____ ____ ____

The prime factorizations of some other numbers follow.

$6 = 2 \cdot 3$

$35 = 5 \cdot 7$ (note $7 \cdot 5$ is the same factorization of 35)

$33 = 3 \cdot 11$

$12 = 2 \cdot 2 \cdot 3$

Remember that each of the factors must be a prime number.

Any prime number may be used as many times as needed. For example:

$$16 = 2 \cdot 2 \cdot 2 \cdot 2$$

Mathematicians have a shortcut for writing $2 \cdot 2 \cdot 2 \cdot 2$. They use 2^4 to mean that 2 is used as a factor four times. Thus:

$$2 \cdot 2 \cdot 2 = 2^3$$

$$5 \cdot 5 \cdot 5 \cdot 5 \cdot 5 \cdot 5 = 5^6$$

$$7 \cdot 7 = 7^2$$

$$8 \cdot 8 \cdot 8 = 8^3$$

The small superscript is called an exponent. The prime factorization

$18 = 2 \cdot 3 \cdot 3$ could be written

$18 = 2 \cdot 3^2$.

We can use whichever notation is better in a given situation.

Here are other examples of prime factorization. Supply the corect numbers in the boxes.

1. $20 = 2 \cdot 2 \cdot 5$ or $2^2 \cdot 5$
2. $55 = 5 \cdot 11$
3. $25 = 5 \cdot 5$ or 5^2
4. $51 = 3 \cdot 17$
5. $128 = 2 \cdot 2 \cdot 2 \cdot 2 \cdot 2 \cdot 2 \cdot 2$ or 2^7
6. $27 = 3^3$
7. $125 = 5^3$
8. $216 = 2^3 \cdot 3^3$
9. $\square = 7^3$
10. $\square = 2^6$
11. $\square = 11^3$
12. $\square = 2^3 \cdot 3^2$
13. $\square = 2^2 \cdot 3^2 \cdot 5^2$
14. $13 = \square$
15. $39 = \boxed{3 \cdot 13}$
16. $5 = \boxed{5}$
17. $8 = \square$
18. $11 = \square$
19. $17 = \boxed{17}$
20. $40 = 2^3 \cdot \square$

EXERCISES

Answer True or False.

1. $5^2 = 25$ T
2. $6^3 = 216$ ________
3. $2^4 = 32$ ________
4. $5^2 \cdot 3 = 30$ ________
5. $7^2 = 49$ ________
6. $B^3 = B \cdot B \cdot B$ ________
7. $2^3 \cdot 3^3 = 54$ ________
8. $W^4 = W \cdot W \cdot W \cdot W$ ________
9. $2^{10} = 1024$ ________
10. $2^{10} = 20$ ________
11. $2^2 \cdot 4^2 = 2 \cdot 2 \cdot 2 \cdot 2 \cdot 2 \cdot 2$ ________
12. $6^3 \cdot 2^2 = 6 \cdot 6 \cdot 6 \cdot 2 \cdot 2$ ________
13. $1^4 = 1$ ________
14. $0^8 = 0$ ________
15. $(14 \cdot 3)^2 = (14 \cdot 3)\ (14 \cdot 3)$ ________
16. $2 \cdot 2 \cdot 3 \cdot 2 = 2^3 \cdot 3$ ________
17. $3 \cdot 5 \cdot 3 \cdot 5 = 3^2 \cdot 5$ ________
18. $15 \cdot 3 = 5 \cdot 3^3$ ________
19. $3^2 \cdot 2^2 = (3 \cdot 2)^2$ ________
20. $3^3 \cdot 2^2 = 6^5$ ________
21. $25 \cdot 9 = (5 \cdot 3)^2$ ________
22. $2^8 \cdot 2^2 = 1024$ ________

METHODS FOR DETERMINING PRIME FACTORS

Example I: Determine the prime factorization of 48.

First factor 48 in any way you please. The factors need not be prime. Thus

$48 = 8 \cdot 6$

$48 = 4 \cdot 2 \cdot 3 \cdot 2$ Next factor 8 into any 2 factors and then 6.

$48 = 2 \cdot 2 \cdot 2 \cdot 3 \cdot 2$ Next factor any of the factors which are not prime.

or

$48 = 2^4 \cdot 3$ Answer

Example II: Determine the prime factorization of 108.

$108 = 12 \quad 9$

$= 3 \cdot 4 \cdot 3 \cdot 3$

$= 3 \cdot 2 \cdot 2 \cdot 3 \cdot 3$ At this step all factors are prime.

Thus $108 = 2^2 \cdot 3^3$

The above examples illustrate the "tree method." Use this method to find the prime factorization of the following numbers. Some are partially done.

Problem I: $540 = 9 \cdot$ ____

$= 3 \cdot$ ____ $\cdot$ 12 $\cdot$ ____

$= 3 \cdot$ ____ $\cdot$ 3 $\cdot$ ____ $\cdot$ ____

$= 3 \cdot$ ____ $\cdot$ 3 $\cdot$ ____ $\cdot$ ____ $\cdot$ ____

Your answer should indicate that $540 = 2^2 \cdot 3^3 \cdot 5$.

Problem 2: 882 = [9] · [ ]

=

=

=

Answer: $882 = 2 \cdot 3^2 \cdot 7^2$

Problem 3: 36 =

Answer: 36 =

Problem 4: 72 =

Answer: 72 =

Problem 5: 12 =

Answer: 12 =

Problem 6: 16 =

Answer: 16 =

Another method for determining the prime factorization of a number is by repeated division by prime numbers.

For instance, the prime factorization of 48 could begin with a division by 2 in the following manner:

Example 1: $2\,\underline{|\,48}$

The 24 can be divided by 2 again. $2\,\underline{|\,24}$

Now the 12 can be divided by 2. $2\,\underline{|\,12}$

6 can also be divided by 2. $2\,\underline{|\,6}$

3 is prime therefore the process stops. 3

Thus the prime factorization of 48 is: $2 \cdot 2 \cdot 2 \cdot 2 \cdot 3$ or $2^4 \cdot 3$.

Notice that each of the divisors is a prime number. The process stops when the dividend is a prime number.

Example 2: Find the prime factorization of 108.

$$\begin{array}{r|l} 3 & \underline{108} \\ 2 & \underline{36} \\ 2 & \underline{18} \\ 3 & \underline{9} \\ & 3 \end{array}$$

Therefore the prime factorization of 108 is:

$3 \cdot 2 \cdot 2 \cdot 3 \cdot 3$ or $2^2 \cdot 3^3$

Problem 1: Using the foregoing process fill in the boxes.

$$\begin{array}{r|l} \square & \underline{12} \\ 3 & \underline{6} \\ & \square \end{array}$$

The prime factorization of 12 is $2^2 \cdot 3$.

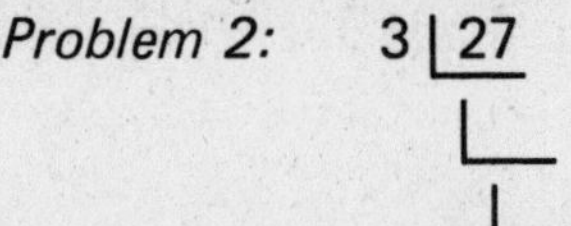

Problem 2:

The prime factorization of 27 is

________________.

Problem 3: 882

882 = ________________

The factorization of 882 by the method of "repeated division" will give the same results as those obtained by using the "tree" method on page 55. This is true of *all* numbers. That is, every number can be factored into a product of *prime* factors in one and only one way.

EXERCISES

Find the prime factorization of each of the following.

1. 18 = ________________
2. 39 = ________________
3. 60 = ________________
4. 100 = ________________

5. 122 = ____________________

6. 47 = ____________________

7. 136 = ____________________

8. 919 = ____________________

9. 440 = ____________________

10. 1024 = ____________________

11. 616 = ____________________

12. 3455 = ____________________

Note: 691 is prime. This can be determined by attempting to divide 691 by all prime numbers that are less than 27. Since 27^2 is greater than 691, prime numbers larger than 27 will yield a quotient less than 27, and these numbers have already been attempted as in the first paragraph.

GREATEST COMMON DIVISOR

In the first part of this chapter divisors of a given number were shown to be all of the counting numbers that will evenly divide the given number.

The counting numbers that are divisors of 60 are:

1, 2, 3, 4, 5, 6, 10, 12, 15, 20, 30 and 60

The divisors of 70 are:

1, 2, 5, 7, 10, 14, 35 and 70

The underlined numbers are *common* divisors. That is, they divide (evenly) both 60 and 70. The common divisors are 1, 2, 5, and 10, and the largest of these, the 10, is called the *Greatest Common Divisor.*

Very frequently the greatest common divisor (which is abbreviated GCD) can be found by inspection.

Consider the pair of numbers 12 and 18. The GCD is 6.

By inspection find the GCD of the following sets of numbers.

1. 6 and 4 ______
2. 3 and 1 ______
3. 7 and 21 ______
4. 4 and 8 ______
5. 12 and 8 ______
6. 21 and 6 ______
7. 4 and 18 ______
8. 50 and 75 ______
9. 6 and 8 and 10 ______
10. ab and ac ______

A very systematic method for determining the GCD of two numbers is to first write the two numbers as a product of primes.

Thus: $60 = 2^2 \cdot 3 \cdot 5$

$70 = 2 \cdot 5 \cdot 7$

Then the GCD is found by selecting all primes that occur in both prime factorizations. In the foregoing case 5 and 2 are primes common to both, and the GCD is the product of these. Therefore the GCD of 60 and 70 is $5 \cdot 2$ or 10.

Find the GCD of 100 and 72.

$$100 = 2 \cdot 2 \cdot 5 \cdot 5 \quad \text{or} \quad 2^2 \cdot 5^2$$
$$72 = 2 \cdot 2 \cdot 2 \cdot 3 \cdot 3 \quad \text{or} \quad 2^3 \cdot 3^2$$

The only numbers that occur in both are $2 \cdot 2$, therefore the GCD is 4.

Write 75 and 45 as a product of prime numbers.

75 = ____ · ____ · ____

45 = ____ · ____ · ____

The primes common to both are ____ and ____.

The GCD is ____.

When a common prime cannot be found we call the two numbers *relatively prime.* In this case the GCD is 1. An example of this would be:

Find the GCD of 15 and 14

$15 = 3 \cdot 5$

$14 = 2 \cdot 7$

There is no common prime factor. These numbers are called ________ ________.

The GCD is _____.

No special trouble is encountered if we wish to find the GCD of 3 numbers. The prime factorization process works very well.

Example: Find the GCD of 140, 78 and 1536.

$140 = 2^2 \cdot 5 \cdot 7$

$78 = 2 \cdot 3 \cdot 13$

$1536 = 2^9 \cdot 3$

The prime number 2 is the only common factor. The GCD is 2.

EXERCISES

Answer True or False.

________ 1. $2^4 = 8$

________ 2. $36 = 3^2 \cdot 2^2$

________ 3. The GCD of 6 and 7 is 1.

________ 4. The GCD of 14 and 16 is 1.

________ 5. The GCD of 27 and 36 is 3.

________ 6. Zero could be a GCD of a pair of numbers.

________ 7. 2 and 4 are relatively prime.

________ 8. Relatively prime numbers must both be prime.

________ 9. $1024 = 2^{10}$

________ 10. $2 \cdot 2 \cdot 2 \cdot 3 \cdot 3 \cdot 5 = 2 \cdot 3 \cdot 2 \cdot 3 \cdot 5 \cdot 2$

Determine the GCD for the following sets of numerals.

Answers

11. 16
    4 __________

12. 32
    20 __________

13. 8
    20 __________

14. 74
    38 __________

15. 24
    84 __________

16. 120
    262 __________

17. 338
    221 __________

18. Mr. Smith has a rectangular garden that measures 48 ft. by 72 ft. He wishes to build a fence around the garden but would like to build it in sections all of equal length.
    (a) Which of these measures could not be the length of each section?

    1 ft 2 ft 3 ft 4 ft 5 ft 6 ft 7 ft

    (b) What is the largest possible measure for the length of each section?

Chapter 6

ADDING AND SUBTRACTING LIKE FRACTIONS

In previous chapters we have studied the four basic operations with counting numbers or whole numbers.

Numbers like $\frac{3}{4}$, $\frac{2}{3}$, and $\frac{7}{4}$ are called fractions. We are now going to work with these numbers. Counting numbers and fractions together are sometimes referred to as "The Numbers of Arithmetic."

Let us look closely at some meanings of fractions in everyday usage.

△ △ ○ △ ○ ○ △ △

____ triangles and ____ circles

____ figures in all

The fractional part of the figures that is triangles is represented by $\frac{5}{8}$ and the part that is circles is represented by $\frac{}{8}$.

This first illustration shows that a fraction might be used to show relationships. $\frac{3}{8}$ can mean 3 parts of 8, or more formally, "the ratio of 3 to 8."

Let's look at another possible meaning of a fraction. Suppose we have a ruler that is marked in units of $\frac{1}{8}$ inch.

In this case $\frac{3}{8}$ means 3 of the units that are each $\frac{1}{8}$ inch. $\frac{11}{8}$ means 11 of the $\frac{1}{8}$ inch units.

A correct definition of the fractions we will use is: "any number that is written as $\frac{a}{b}$ where a and b are counting numbers and b is not 0."

Another meaning of a fraction that we will use later is that $\frac{a}{b}$ means a ÷ b. Now let's look at addition and subtraction of these new numbers.

Suppose we wish to add $\frac{3}{8}$ and $\frac{4}{8}$. Let's think of these fractions as 3 units $\frac{1}{8}$ inch long and 4 units $\frac{1}{8}$ inch long. Obviously the sum is 7 units $\frac{1}{8}$ inch long, which can be called $\frac{7}{8}$. So $\frac{3}{8} + \frac{4}{8} = \frac{7}{8}$

Fractions are said to be *like fractions* if they have the same denominator (bottom number). To add like fractions, add the numerators (top numbers) and place this sum over the common denominator.

Examples: $\frac{4}{11} + \frac{3}{11} = \frac{7}{11}$; $\frac{2}{3} + \frac{4}{3} = \frac{6}{3}$

To subtract like fractions, subtract the numerators and place this difference over the common denominator.

Examples: $\frac{4}{11} - \frac{3}{11} = \frac{1}{11}$; $\frac{7}{8} - \frac{2}{8} = \frac{5}{8}$

Perform the indicated operation in each of the following.

1. $\frac{3}{15} + \frac{9}{15} =$

2. $\frac{7}{3} - \frac{2}{3} =$

3. $\frac{4}{5} + \frac{6}{5} =$

4. $\frac{5}{8} + \frac{6}{8} =$

5. $\frac{2}{10} + \frac{6}{10} =$

6. $\frac{7}{6} + \frac{2}{6} =$

7. $\frac{11}{14} - \frac{7}{14} =$

8. $\frac{10}{3} - \frac{7}{3} =$

9. $\frac{4}{5} + \frac{9}{5} =$

10. $\frac{6}{8} - \frac{2}{8} =$

11. $\frac{3}{5} + \frac{4}{5} - \frac{2}{5} =$

12. $\frac{8}{3} + \frac{2}{3} - \frac{6}{3} =$

13. $\frac{5}{8} + \frac{2}{8} + \frac{3}{8} =$

14. $\frac{7}{4} - \frac{2}{4} - \frac{3}{4} =$

15. $\frac{4}{5} - \frac{1}{5} + \frac{3}{5} =$

16. $\frac{4}{23} + \frac{18}{23} - \frac{3}{23} =$

17. $\frac{7}{4} - \frac{3}{4} - \frac{1}{4} =$

18. $\frac{7}{8} + \frac{3}{8} - \frac{1}{8} =$

19. $\frac{7}{4} - \frac{3}{4} - \frac{1}{4} =$

20. $\frac{7}{8} + \frac{3}{8} - \frac{1}{8} =$

21. $\frac{41}{32} + \frac{19}{32} - \frac{28}{32} =$

22. $\frac{16}{17} - \frac{3}{17} - \frac{2}{17} =$

23. $\frac{5}{9} + \frac{12}{9} + \frac{1}{9} =$

24. $\frac{2}{11} + \frac{7}{11} - \frac{9}{11} =$

REVIEW TEST OF CHAPTERS 1 – 6

1. 2,384,061 in words is ____

____

2. The sum of two hundred eighty four, sixty eight, one thousand three, and two hundred two is ____

3. Add:
```
2841
 983
1004
 762
 996
```

4. Subtract:
```
9008
 769
```

5. Multiply:
```
7003
 261
```

6. Find a value for x to make this a true statement:

   $14 - (6 + 3) - 2 = x$

   x = ________

7. Find the prime factorization of:

   (a) 144 = ______________

   (b) 48 = ______________

   (c) 18 = ______________

8. The GCD of 144, 48, and 18 is ________.

9. Find x if $\frac{7}{8} + \frac{2}{8} - \frac{3}{8} + \frac{1}{8} = x$

   x = ________

10. Divide 7841 by 293 and give the answer in the form of a mathematical sentence.

____

Chapter 7

DIFFERENT NAMES FOR FRACTIONS

Four parts of eight shaded

$\frac{4}{8}$

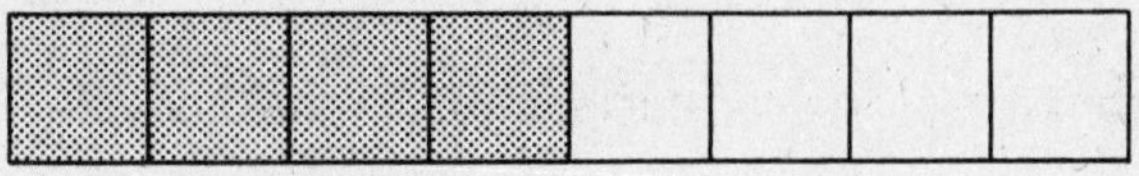

Three parts of six shaded

$\frac{3}{6}$

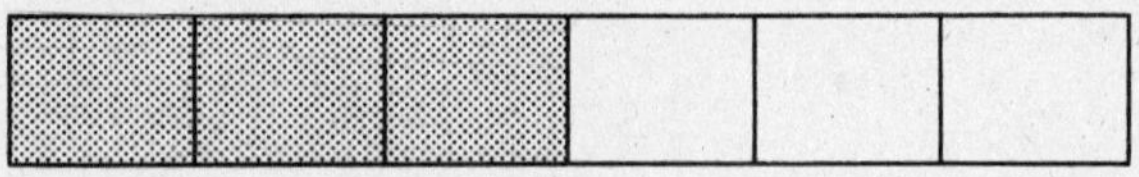

One part of two shaded

$\frac{1}{2}$

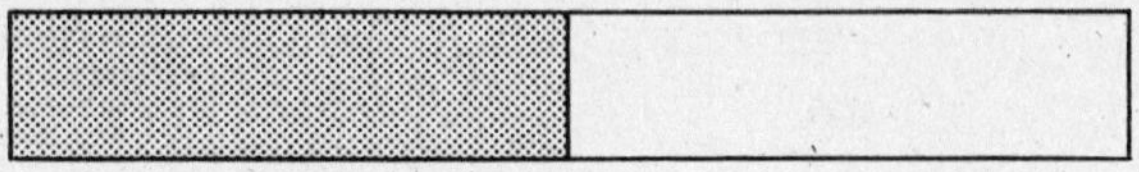

By observing these diagrams it seems obvious that $\frac{1}{2}, \frac{3}{6}$, and $\frac{4}{8}$ all represent the same amount. Actually this is true, and we say that $\frac{1}{2}, \frac{3}{6}$, and $\frac{4}{8}$ are different names for the same number.

Notice that if we multiply both the numerator and denominator of $\frac{1}{2}$ by 3, i.e., $\frac{3 \times 1}{3 \times 2} = \frac{3}{6}$, and by 4, i.e., $\frac{4 \times 1}{4 \times 2} = \frac{4}{8}$, we get the fractions $\frac{3}{6}$ and $\frac{4}{8}$. This should indicate a method of finding different names for fractions. "If both the numerator and the denominator of a fraction are multiplied by any number except 0, the result will be a fraction equal to the original fraction."

EXERCISES

Give three other fractions equal to each of the given fractions.

1. $\frac{2}{3}$ $\frac{4}{6}$, $\frac{8}{12}$, $\frac{16}{24}$
2. $\frac{3}{4}$ ____, ____, ____
3. $\frac{5}{8}$ $\frac{10}{16}$, $\frac{20}{32}$, $\frac{40}{64}$
4. $\frac{1}{4}$ ____, ____, ____
5. $\frac{2}{6}$ ____, ____, ____
6. $\frac{1}{3}$ ____, ____, ____
7. $\frac{1}{5}$ ____, ____, ____
8. $\frac{2}{10}$ ____, ____, ____

We should also note that if we have a fraction like $\frac{4}{8}$ we can divide the numerator and denominator by 4, i.e., $\frac{4 \div 4}{8 \div 4}$, and get $\frac{1}{2}$, which is equal to $\frac{4}{8}$. This leads us to another rule. "If both numerator and denominator of a given fraction are divided by the same number, the resulting fraction is equal to the given fraction."

Examples:

$$\frac{5}{15} = \frac{5 \div 5}{15 \div 5} = \frac{\ }{3}$$

$$\frac{6}{8} = \frac{6 \div 2}{8 \div 2} = \frac{\ }{4}$$

A fraction is said to be in "simplest form" or "lowest terms" when there is no prime number that will divide both numerator and denominator.

How do we determine if a fraction is in simplest form?

The answer to this question is really quite simple. Remember that in Chapter 9 we learned that any number can be factored into primes. If we factor both the numerator and denominator into primes, then we can observe any numbers that will divide both.

Example: Find the simplest form of the fraction $\frac{126}{198}$.

$\frac{126}{198} = \frac{2 \times 3 \times 3 \times 7}{2 \times 3 \times 3 \times 11}$ We can see that the numerator 126 and the denominator 198 can be divided by $2 \times 3 \times 3$.

$\frac{126}{198} = \frac{2 \times 3 \times 3 \times 7}{2 \times 3 \times 3 \times 11} = \frac{7}{11}$ So $\frac{7}{11}$ is the simplest form of $\frac{126}{198}$.

Let's state a rule for simplifying fractions. "To find the simplest form of a fraction, factor the numerator and denominator into primes and divide each by all of the common primes."

EXERCISES

Give the simplest form of each of the following.

1. $\frac{8}{12} = \frac{\ \times\ \times\ }{\ \times\ \times\ } =$ __________

2. $\frac{21}{63} =$ __________

3. $\frac{18}{48} =$ __________

4. $\frac{7}{42} =$ __________

5. $\frac{9}{27} =$ __________

6. $\frac{81}{121} =$ __________

7. $\frac{69}{91} =$ __________

8. $\frac{126}{212} =$ __________

9. $\frac{84}{116} =$ __________

10. $\frac{14}{32} + \frac{8}{32} = \frac{}{32} =$ __________

11. $\frac{17}{52} - \frac{15}{52} = \frac{}{52} =$ __________

12. $\frac{14}{8} - \frac{6}{8} = \frac{}{8} =$ __________

Find the missing numbers to make the following sentences true.

13. $\frac{7}{8} = \frac{}{24}$

14. $\frac{9}{12} = \frac{}{36}$

15. $\frac{2}{5} = \frac{}{5 \times 3 \times 7}$

16. $\frac{21}{36} = \frac{}{144}$

17. $\frac{9}{15} = \frac{}{90}$

18. $\frac{25}{8} = \frac{}{1000}$

19. $\frac{7}{9} = \frac{}{180}$

20. $\frac{4}{11} = \frac{}{88}$

21. $\frac{7}{16} - \frac{3}{16} = \frac{}{16} = \frac{}{4}$

22. $\frac{2}{3} + \frac{2}{3} = \frac{}{3} = \frac{}{12}$

MIXED NUMBERS

A "mixed number" is a number like $3\frac{4}{5}$. It is called a mixed number because it is part whole number and part fraction.

$3\frac{4}{5}$ actually means $3 + \frac{4}{5}$. We read it as "three and four fifths" and agree to leave out the plus sign when we write it.

Basically we have two problems involving mixed numbers. We may wish to change a mixed number to a fraction, or a fraction to a mixed number.

To change $3\frac{4}{5}$ to a fraction we think of it as $3 + \frac{4}{5}$. Other names for 3 might be $\frac{6}{2}, \frac{9}{3}, \frac{\_\_}{4}, \frac{\_\_}{5}$, and so on. The one in which we are interested is $\frac{15}{5}$ since $3 + \frac{4}{5}$ can now be written as $\frac{15}{5} + \frac{4}{5}$.

So $3\frac{4}{5} = \frac{15}{5} + \frac{4}{5} = \frac{\_\_}{5}$

$$2\frac{3}{4} = 2 + \frac{3}{4} = \frac{\_\_}{4} + \frac{3}{4} = \frac{\_\_}{4}$$

$$1\frac{1}{7} = 1 + \frac{1}{7} = \frac{\_\_}{7} + \frac{1}{7} = \frac{\_\_}{7}$$

We soon see that our whole number should be renamed, using the denominator of the fractional part. The easy way to change $3\frac{4}{5}$ to a fraction is to think $3\frac{4}{5} = \frac{(5 \times 3) + 4}{5} = \frac{\_\_}{5}$.

Also $2\frac{3}{4} = \frac{(4 \times 2) + 3}{4}$

We can state a rule for changing mixed numbers to fractions. "Multiply the denominator of the fractional part by the whole number and add to this the numerator of the fractional part. This gives us our numerator which is placed over the denominator of the fractional part."

Example: $5\frac{7}{8}$; $8 \times 5 = 40$; $40 + 7 = 47$

So $5\frac{7}{8} = \frac{47}{8}$

Change the following mixed numbers to fractions.

1. $5\frac{7}{8} = \frac{\quad}{8}$

2. $4\frac{1}{5} = \frac{\quad}{5}$

3. $16\frac{1}{2} = \frac{\quad}{2}$

4. $8\frac{2}{3} = \frac{\quad}{3}$

5. $9\frac{6}{11} = \frac{\quad}{11}$

6. $6\frac{\quad}{10} = \frac{\quad}{10}$

7. $9\frac{4}{5} = \frac{\quad}{5}$

8. $21\frac{1}{4} = \frac{\quad}{4}$

9. $15\frac{3}{8} = \frac{\quad}{8}$

10. $26\frac{1}{3} = \frac{\quad}{3}$

The problem of changing a fraction to a mixed number will occur only when the numerator is larger than the denominator. Historically such fractions are called "improper fractions." Unit fractions (those with a value of 1) are also in the set of improper fractions. For instance $\frac{16}{3}$ is an improper fraction; so is $\frac{3}{3}$.

Let's change $\frac{16}{3}$ to a mixed number. Actually we break $\frac{16}{3}$ into two parts.

$$\frac{16}{3} = \frac{15}{3} + \frac{1}{3} = 5 + \frac{1}{3} = 5\frac{1}{3}.$$

15 is the largest multiple of 3 that is less than 16. This change can be accomplished by dividing 3 into 16 and placing the remainder as the numerator of the fractional part. $16 \div 3 =$ 5 with a remainder of 1 or $\frac{16}{3} = 5\frac{1}{3}$.

Example: Change $\frac{25}{2}$ to a mixed number.

$25 \div 2 = 12$ with remainder 1

$\frac{25}{2} = 12\frac{1}{2}$

Example: Change $\frac{38}{6}$ to a mixed number.

$38 \div 6 = 6$ with a remainder of 2

$\frac{38}{6} = 6\frac{2}{6}$; But $\frac{2}{6}$ is not in simplest form.

$\frac{2}{6} = \frac{2}{2 \times 3} = \frac{1}{3}$

So $\frac{38}{6} = 6\frac{1}{3}$ in simplest form.

EXERCISES

Change each improper fraction to a mixed number.

1. $\frac{7}{2}$ = __________

2. $\frac{53}{4}$ = __________

3. $\frac{17}{3}$ = __________

4. $\frac{29}{7}$ = __________

5. $\frac{112}{10}$ = __________

6. $\frac{78}{5}$ = __________

7. $\frac{64}{9}$ = __________

8. $\frac{47}{12}$ = __________

9. $\frac{15}{2}$ = __________

10. $\frac{114}{11}$ = __________

Chapter 8
MULTIPLICATION OF FRACTIONS

One third or the students in a class of 30 made "C" on the first test. How many students made "C"?

Our problem here is to find $\frac{1}{3}$ of 30 or the product of $\frac{1}{3}$ and 30.

Remember that $\frac{1}{3}$ can mean 1 part of 3. In our problem it would mean 1 student of every 3.

To multiply $\frac{1}{3}$ times 30 we are really asking "how many groups of 3 are contained in a group of 30?"

$$\frac{1}{3} \times 30 = \frac{1 \times 30}{3} = \frac{30}{3} = \_\_\_\_$$

So 10 students made a grade of "C".

Suppose the problem had stated that $\frac{2}{3}$ of the students made "C".

We now have $\frac{2}{3} \times 30$ as a problem.

$$\frac{2}{3} \times 30 = \frac{2 \times 30}{3} = \frac{\_\_}{3} = \_\_\_\_$$

"To multiply a fraction by a whole number, multiply the numerator of the fraction by the whole number. This gives the numerator of the product, which is placed over the denominator of the original fraction."

Example: $\frac{3}{4} \times 11 = \frac{3 \times 11}{4} = \frac{\_\_}{4}$

The product of $\frac{3}{4}$ and 11 may now be changed to a mixed number.

$$\frac{3}{4} \times 11 = \frac{3 \times 11}{4} = \frac{\_\_}{4} = \_\_\_\_$$

EXERCISES

Perform the multiplication and change the answers to mixed numbers.

1. $\frac{5}{8} \times 40 =$ ________

2. $\frac{2}{5} \times 33 = \frac{\_\_}{5} =$ ________

3. $\frac{7}{8} \times 16 =$ ________

4. $\frac{2}{3} \times 26 = \frac{\_\_}{3} =$ ________

5. $\frac{9}{5} \times 80 =$ ________

6. $\frac{7}{3} \times 30 =$ ________

7. $\frac{1}{16} \times 60 = \frac{\_\_}{16} =$ ________

8. $\frac{9}{11} \times 40 = \frac{\_\_}{11} =$ ________

9. $\frac{4}{3} \times 12 =$ ________

10. $\frac{9}{16} \times 84 = \frac{\_\_}{16} =$ ________

Next we will consider the product of two fractions.

Example: $\frac{2}{3} \times \frac{3}{4}$

Our task in this example is to find $\frac{2}{3}$ of $\frac{3}{4}$. We might picture the problem in this way.

3 parts of 4 or $\frac{3}{4}$ of the bar is shaded. If we take the $\frac{3}{4}$ that is shaded and divide it into 3 parts, then $\frac{2}{3}$ of $\frac{3}{4}$ will be represented by 2 of these parts.

The darker shaded part represents $\frac{2}{3}$ of the $\frac{3}{4}$ that is shaded. We can see that this is $\frac{1}{2}$ of the original bar.

So $\frac{2}{3} \times \frac{3}{4} = \frac{1}{2}$

We can discover a rule from this example by observing that

$$\frac{2}{3} \times \frac{3}{4} = \frac{2 \times 3}{3 \times 4} = \frac{6}{12} = \frac{1}{2}$$

"The product of two or more fractions is a fraction whose numerator is the product of the numerators and whose denominator is the product of the denominators."

$$\frac{a}{b} \times \frac{c}{d} = \frac{a \times c}{b \times d}$$

Example: $\frac{5}{7} \times \frac{2}{3} = \frac{5 \times 2}{7 \times 3} = \frac{\quad}{21}$

Example: $\frac{2}{9} \times \frac{5}{7} = \frac{5 \times 2}{9 \times 7} =$ ________

EXERCISES

Find the following products. Leave your answers as fractions in simplest form (i.e., numerator and denominator have no common divisor).

1. $\frac{5}{8} \times \frac{2}{5} =$ ________

2. $\frac{1}{2} \times \frac{4}{3} =$ ________

3. $\frac{7}{16} \times \frac{4}{5} =$ ________

4. $\frac{9}{11} \times \frac{77}{84} =$ ________

5. $\frac{1}{7} \times \frac{2}{10} =$ ________

6. $\frac{4}{11} \times \frac{3}{4} =$ ________

7. $\frac{9}{16} \times \frac{7}{8} =$ ________

8. $\frac{6}{13} \times \frac{26}{36} =$ ________

9. $\frac{2}{3} \times \frac{16}{19} =$ ________

10. $\frac{2}{3} \times \frac{4}{7} =$ ________

To multiply a mixed number by a fraction, we first change the mixed number to an improper fraction.

Example: $$4\frac{2}{3} \times \frac{9}{12} = \frac{14}{3} \times \frac{9}{12}$$

$$= \frac{14 \times 9}{3 \times 12} = \frac{}{36} = 3\frac{}{36} = 3\frac{1}{2}$$

Let's look closely at the problem of changing the answer to simplest form. Above we get $\frac{126}{36}$, which must be reduced to $3\frac{1}{2}$. From our study of how to reduce fractions you should remember that the numerator and denominator must be factored and then divided by their greatest common divisor.

This leads us to a shortcut that is very useful. We can actually reduce our answer before we get it and make our work easier at the same time.

Observe that in the foregoing example, $\frac{14}{3} \times \frac{9}{12}$, we multiplied 14 x 9 and 3 x 12 and then factored them. If we realize that these products are already partially factored we can save some work.

$$\frac{14}{3} \times \frac{9}{12} = \frac{14 \times 9}{3 \times 12} = \frac{7 \times \overset{1}{\cancel{2}} \times \overset{1}{\cancel{3}} \times \overset{1}{\cancel{3}}}{\underset{1}{\cancel{3}} \times \underset{1}{\cancel{3}} \times \underset{1}{\cancel{2}} \times 2} = \frac{7}{2} = 3\frac{1}{2}$$

Remember that we can divide numerator and denominator by the same number and preserve the value of the fraction.

Example: $$\frac{\overset{1}{\cancel{3}}}{\underset{1}{\cancel{4}}} \times \frac{\overset{3}{\cancel{12}}}{\underset{5}{\cancel{15}}} = \frac{1 \times 3}{1 \times 5} = \frac{3}{5}$$

The numerator of the first fraction and denominator of the second are each divided by 3 and the numerator of the second fraction and denominator of the first are each divided by 4.

Example: $$\frac{\overset{1}{\cancel{7}}}{\underset{2}{\cancel{8}}} \times \frac{\overset{\overset{1}{3}}{\cancel{12}}}{\underset{\underset{1}{3}}{\cancel{21}}} = \frac{1 \times 1}{2 \times 1} = \frac{1}{2}$$

In this example we can see that this division is not necessarily from the numerator of one fraction to the denominator of another.

Example: $$\frac{\overset{1}{\cancel{4}}}{\underset{2}{\cancel{8}}} \times \frac{3}{7} = \frac{1 \times 3}{2 \times 7} = \frac{3}{14}$$

Also: $$\frac{\overset{1}{\cancel{2}}}{\underset{1}{\cancel{3}}} \times \frac{\overset{1}{\cancel{5}}}{\underset{4}{\cancel{8}}} \times \frac{\overset{1}{\cancel{3}}}{\underset{1}{\cancel{5}}} = \frac{1 \times 1 \times 1}{1 \times 4 \times 1} = \frac{1}{4}$$

EXERCISES

Find the products. Answers should be left as fractions in simplest form.

1. $\frac{3}{7} \times \frac{8}{11} =$

2. $2\frac{1}{2} \times 20\frac{1}{5} =$

3. $\frac{3}{7} \times \frac{8}{12} =$

4. $\frac{2}{3} \times \frac{1}{2} \times \frac{3}{5} =$

5. $5\frac{1}{3} \times \frac{3}{4} =$

6. $\frac{5}{8} \times 24 =$

7. $\frac{2}{3} \times 8 \times \frac{3}{4} =$

8. $4\frac{1}{2} \times 2\frac{2}{3} =$

9. $\frac{15}{2} \quad \frac{8}{12} =$

10. $7\frac{1}{2} \times \frac{2}{3} =$

11. $6\frac{1}{4} \times \frac{2}{5} \times 3\frac{5}{8} =$

12. $16\frac{1}{2} \times \frac{2}{3} =$

13. $\frac{7}{1} \times \frac{2}{5} \times \frac{3}{7} =$

14. $\frac{2}{3} \times \frac{1}{4} \times 6 =$ __________

15. $5\frac{1}{4} \times \frac{2}{7} \times \frac{2}{3} =$ __________

16. $\frac{1}{2} \times \frac{1}{3} \times \frac{1}{5} =$ __________

17. A coat is on sale and marked "$\frac{1}{3}$ off." If the original price was $24, what is the sale price? $16.00

18. In a certain city a survey shows that $\frac{7}{9}$ of the homes have television sets. If there are 18,972 homes in the city, how many have television sets? 14,756

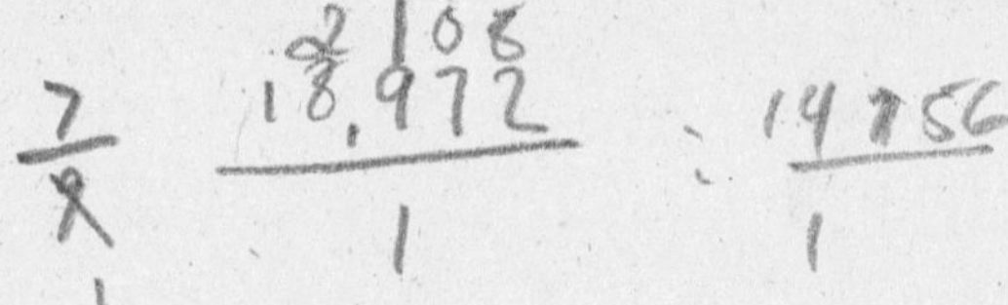

19. John has read $\frac{2}{3}$ of a novel. If there are 861 pages, how many more pages must he read to finish the book? 287

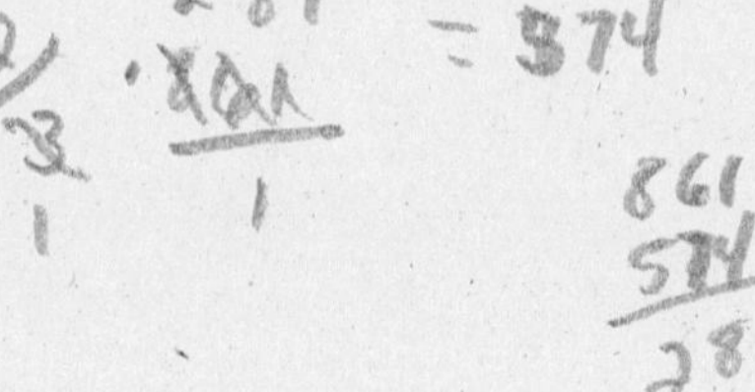

20. Mr. Jones gives his wife $\frac{1}{2}$ of his salary for household expenses. She spends $\frac{2}{3}$ of this for groceries. Last week Mr. Jones made $126. How much did Mrs. Jones spend for groceries. $42.00

EXERCISES

Leave answers as proper fractions in simplest form or as mixed numbers.

1. $\frac{5}{8} \times \frac{3}{4} \times \frac{8}{5} =$ __________

2. $728 \times \frac{3}{8}$

3. $67\frac{1}{2} \times \frac{3}{5} \times \frac{2}{3} =$ __________

4. $87\frac{1}{2} \times 100 =$ __________

5. $62\frac{1}{2} \times 7\frac{1}{2} =$ __________

6. $1\frac{2}{3} \times \frac{3}{5} \times 7\frac{1}{2} \times \frac{2}{15} =$ __________

7. $24\frac{1}{2} \times \frac{3}{7} \times \frac{2}{3} =$ __________

8. $826 \times 4\frac{1}{2} =$ __________

9. $1864 \times 2\frac{1}{3} =$ __________

10. $\frac{2}{3} \times \frac{7}{8} \times \frac{4}{5} \times 1\frac{5}{8} \times \frac{3}{7} =$ __________

Chapter 9

DIVISION OF FRACTIONS

In this chapter we will consider the problems of dividing a whole number by a fraction, a fraction by a whole number, and a fraction by a fraction.

Remember our interpretation of division.

$4 \div \frac{1}{2}$ asks "how many one halves are contained in 4?"

$\frac{1}{2} \div 4$ asks "how many fours are contained in $\frac{1}{2}$?"

$\frac{1}{3} \div \frac{3}{4}$ asks "how many three-fourths are contained in $\frac{1}{3}$?"

Let's first look at the problem $4 \div \frac{1}{2}$. We have already seen that division can be indicated more than one way. We will prefer to write $4 \div \frac{1}{2}$ as $\frac{4}{\frac{1}{2}}$. This expression is known as a *complex fraction.* A fraction is said to be complex if the numerator or denominator or both are themselves fractions.

Recall that one property of a fraction is that its value is unchanged if numerator and denominator are multiplied by the same non-zero number. Such a multiplication is in truth multiplying by 1 since $\frac{n}{n} = 1$ for any n that is not zero.

Now let's apply the property of multiplying by 1 to the complex fraction $\frac{4}{\frac{1}{2}}$. Suppose we decide to multiply the numerator and denominator by 5. We will get $\frac{4 \times 5}{\frac{1}{2} \times 5}$. Since this does not change the value of the fraction we can say that $\frac{4}{\frac{1}{2}} = \frac{20}{\frac{5}{2}}$. This same type of statement will be true regardless of the number we use as a multiplier. However, multiplying by 5 did nothing except give us a more complicated expression, and our goal is to simplify the expression.

Therefore we need to multiply by a number that will make the complex fraction a simple fraction. 2 is such a number since $\frac{4 \times 2}{\frac{1}{2} \times 2} = \frac{8}{1} = 8$; so $4 \div \frac{1}{2} = 8$

Examples: (a) $7 \div \frac{2}{3}$

$$\frac{7}{\frac{2}{3}} = \frac{7 \times \frac{3}{2}}{\frac{2}{3} \times \frac{3}{2}} = \frac{\frac{21}{2}}{1} = \frac{21}{2} = 10\frac{1}{2}$$

So $7 \div \frac{2}{3} = 10\frac{1}{2}$

(b) $9 \div \frac{4}{5}$

$$\frac{9}{\frac{4}{5}} = \frac{9 \times \frac{5}{4}}{\frac{4}{5} \times \frac{5}{4}} = \frac{\frac{45}{4}}{1} = \frac{45}{4} = 11\frac{1}{4}$$

So $9 \div \frac{4}{5} = 11\frac{1}{4}$

Notice in these examples that a multiplier is chosen that will give us a 1 in the denominator. Can such a number always be found?

EXERCISES

Fill in the blanks.

1. $\frac{2}{3} \times$ ____ $= 1$

2. $\frac{7}{8} \times$ ____ $= 1$

3. $4 \times$ ____ $= 1$

4. $2\frac{1}{2} \times$ ____ $= 1$

5. $\frac{8}{6} \times$ ____ $= 1$

6. $5 \times$ ____ $= 1$

Note: If A × B = 1 then A and B are said to be reciprocals of each other. In problem 3, four is the reciprocal of ¼ and ¼ is the reciprocal of 4. Each number, except 0, has a reciprocal.

EXERCISES

To solve these problems, first write the indicated division as complex fraction and then find a multiplier that will give a denominator of 1.

1. $6 \div \frac{2}{3}$

2. $18 \div \frac{4}{5}$

3. $12 \div \frac{1}{3}$

4. $15 \div 6\frac{1}{2}$

5. $8 \div 2\frac{1}{3}$

6. $10 \div \frac{1}{2}$

7. $9 \div 2\frac{1}{3}$

8. $12 \div 4\frac{2}{3}$

9. $7 \div \frac{3}{7}$

10. $15 \div \frac{1}{4}$

11. $54 \div \frac{1}{3}$

12. $100 \div \frac{4}{5}$

13. $16 \div 2\frac{3}{4}$

14. $25 \div 4\frac{1}{5}$

15. $25 \div 12\frac{1}{2}$

16. $26 \div 3\frac{7}{8}$

17. $9 \div 12\frac{1}{2}$

Let's now consider the problem of a fraction divided by a fraction.

$$\frac{2}{3} \div \frac{3}{4}$$

Our complex fraction in this case is $\dfrac{\frac{2}{3}}{\frac{3}{4}}$

Once again we will simplify this complex fraction by multiplying numerator and denominator by the same number — a number that will give us a 1 in the denominator.

$$\frac{2}{3} \div \frac{3}{4} = \frac{\frac{2}{3}}{\frac{3}{4}} = \frac{\frac{2}{3} \times \frac{4}{3}}{\frac{3}{4} \times \frac{4}{3}} = \frac{\frac{8}{9}}{1} = \frac{8}{9}$$

So $\frac{2}{3} \div \frac{3}{4} = \frac{8}{9}$

Example: $4\frac{1}{2} \div 2\frac{1}{3} = \dfrac{4\frac{1}{2}}{2\frac{1}{3}} = \dfrac{\frac{9}{2}}{\frac{7}{3}} = \dfrac{\frac{9}{2} \times \frac{3}{7}}{\frac{7}{3} \times \frac{3}{7}} = \dfrac{\frac{27}{14}}{1} = \frac{27}{14} = 1\frac{13}{14}$

So $4\frac{1}{2} \div 2\frac{1}{3} = 1\frac{13}{14}$

EXERCISES

Write these as complex fractions then solve.

1. $3\frac{1}{2} \div \frac{2}{3}$

2. $\frac{3}{5} \div \frac{3}{4}$

3. $1\frac{1}{2} \div \frac{4}{5}$

4. $3\frac{3}{4} \div 7$

5. $9\frac{1}{2} \div 2\frac{2}{3}$

6. $4\frac{1}{4} \div 3\frac{1}{2}$

7. $\frac{14}{3} \div \frac{3}{14}$

8. $\frac{12}{5} \div 5\frac{1}{2}$

9. $\frac{17}{3} \div 5\frac{2}{3}$

10. $\frac{1}{2} \div \frac{7}{8}$

The method of writing division problems as complex fractions should give rise to a rule for dividing fractions.

Example: $\frac{2}{3} \div \frac{3}{4}$

$$\frac{2}{3} \div \frac{3}{4} = \frac{\frac{2}{3}}{\frac{3}{4}} = \frac{\frac{2}{3} \times \frac{4}{3}}{\frac{3}{4} \times \frac{4}{3}} = \frac{\frac{2}{3} \times \frac{4}{3}}{1} = \frac{2}{3} \times \frac{4}{3}$$

So $\frac{2}{3} \div \frac{3}{4} = \frac{2}{3} \times \frac{4}{3} =$ ________.

Example: $\frac{3}{7} \div \frac{5}{8}$

$$\frac{3}{7} \div \frac{5}{8} = \frac{\frac{3}{7}}{\frac{5}{8}} = \frac{\frac{3}{7} \times \frac{8}{5}}{\frac{5}{8} \times \frac{8}{5}} = \frac{\frac{3}{7} \times \frac{8}{5}}{1} = \frac{3}{7} \times \frac{8}{5}$$

So $\frac{3}{7} \div \frac{5}{8} = \frac{3}{7} \times \frac{8}{5} =$ ________.

The last line of each example shows the division problem equal to a multiplication problem.

Write a rule for division of fractions.

____

____

____

____

____

____

EXERCISES

Use the rule you wrote on the previous page to solve these problems.

1. $\frac{7}{8} \div 4\frac{1}{2}$

2. $\frac{3}{4} \div \frac{4}{5}$

3. $5\left(\frac{1}{4} \div \frac{3}{4}\right)$

4. $\left(\frac{2}{3}\right)\left(\frac{7}{8} \div \frac{1}{4}\right)$

5. $4\frac{2}{3} \div \frac{1}{3}$

6. $\frac{7}{16} \div 3\frac{1}{2}$

7. $4\frac{1}{2}x = 12$

$x =$ ________

8. $\frac{2}{3}x = 14$

$x =$ ________

9. $\frac{7}{8}x = 2\frac{1}{2}$

$x =$ ________

10. $\frac{1}{4}x = 3\frac{7}{8}$

$x =$ ________

EXERCISES

1. $(\frac{7}{8} \times \frac{2}{3}) \div 7 =$ ________

2. $\frac{7}{8} \times (\frac{2}{3} \div 7) =$ ________

3. $(\frac{3}{7} + \frac{2}{7}) \times \frac{4}{5} =$ ________

4. $(\frac{3}{7} \times \frac{4}{5}) + (\frac{2}{7} \times \frac{4}{5}) =$ ________

5. $(2\frac{1}{3} \div 6\frac{1}{2}) \times \frac{5}{8} =$ ________

6. $(4\frac{1}{8} \times 1\frac{1}{3}) + (2\frac{5}{8} \times 1\frac{1}{3}) =$ ________

7. $(4\frac{1}{8} + 2\frac{5}{8}) \times 1\frac{1}{3} =$ ________

8. To make a set of drapes, Mrs. Brown needed 14½ yards of drapery material and 14½ yards of lining material. If the drapery material cost $3.12 per yard and the lining cost 92 cents per yard, how much would she pay for all the material?

9. If Mr. Jones drove his car for 77 miles and used 3½ gallons of gasoline, how many miles per gallon did he average?

10. 12 inches = 1 ft. and 3 ft. = 1 yard. How many inches in 14½ yards?

Chapter 10

MULTIPLES AND LEAST COMMON MULTIPLE

In Chapter 6 we added fractions with the same denominator such as $\frac{3}{8} + \frac{2}{8} = \frac{5}{8}$. We now must learn to add fractions such as $\frac{7}{8} + \frac{3}{7}$. Fractions with unlike denominators cannot be combined unless we first change them to equivalent fractions having like denominators. This chapter deals with finding the least common multiple of any set of numbers.

The least common multiple will be abbreviated L. C. M.

First, let's consider the multiples of a number that are obtained by multiplying any given number by 1, then 2, then 3, etc.

For example, the multiples of 2 are 2, 4, 6, 8, etc. The multiples of 6 are 6, 12, 18, 24, etc. Multiples of "a" might be called $1 \times a$, $2 \times a$, $3 \times a$, $4 \times a$, $5 \times a$, etc. if a represents some number.

There is not a last multiple of a number, and this has been indicated by the use of the abbreviation "etc." A more common manner for showing the continuation of a sequence is by the use of three dots. For example, the multiples of 11 are 11, 22, 33, 44, 55, $\cdots$.

EXERCISES

Show multiples of the following:

1. 12: ____, ____, ____, ____, ____, . . .
2. 5: ____, ____, ____, . . .
3. 7: ____, ____, ____, ____, ____, . . .
4. 14: ____, ____, ____, ____, ____, ____, . . .
5. 2041: ____, ____, ____, . . .
6. ½: ____, ____, ____, ____, . . .
7. ¾: ____, ____, ____, ____, ____, . . .

Supply the missing number in the box also.

8. ☐ : ____, 22, ____, 44, ____, ____, . . .
9. ☐ : ____, ____, ____, ____, ____, ____, 49, . . .
10. ☐ : ____, ____, 60, ____, . . .
11. 3: ____, ____, ____, ____, ____, ____, ____, ____, ____, ____, ____, . . .
12. 4: ____, ____, ____, ____, ____, ____, ____, ____, ____, . . .

Notice that the number 12 occurs as a multiple of 3 and as a multiple of 4. (See problems 11 and 12 above.) It is therefore a *Common* multiple of 3 and 4. ("Common" means "possessed by both.") Another common multiple of 3 and 4 can be found in problems 11 and 12. It is ____.

List some other common multiples of 3 and 4. ____, ____, ____. Is 120 a common multiple of 3 and 4? ____.

Common multiples of 5 and 7 include 35, 70, 105, 140, 175, and 210.

Common multiples of 4 and 6 are 12, 24, 36, 48, 60, 72, ... Notice that both 4 and 6 must evenly divide each of these common multiples. Some common multiples of 4 and 5 are ____, ____, ____, ____, ____, ____, ... Common multiples of 6 and 3 are ____, ____, ____, ____, ...

Now, the *least common multiple* is (as you have already guessed) the smallest of the common multiples of any two numbers.

The least common multiple (LCM) of 3 and 4 is 12. The LCM of 5 and 7 is 35. The L.C.M. of 4 and 6 is 12. What have you found (see previous exercises) to be the LCM of 4 and 5? ____ Of 6 and 3? ____.

Find the L.C.M of the following pairs of numbers:

1. 2 and 9 ________
2. 3 and 10 ________
3. 4 and 11 ________
4. 4 and 12 ________
5. 3 and 11 ________
6. 6 and 8 ________
7. 6 and 13 ________
8. $\frac{1}{6}$ and $\frac{1}{8}$ ________
9. 7 and 3 ________
10. 11 and 13 ________

By using the same process we can find the LCM of three numbers. For example, the L.C.M. of 3, 6 and 10 can be found by writing some multiples of each and finding the smallest of the common multiples.

3: 3, 6, 9, 12, 15, 18, 21, 24, 27, 30, 33

6: 6, 12, 18, 24, 30, 36, 42, 48

10: 10, 20, 30, 40, 50, 60, 70, 80, 90

In this case, the LCM is 30.

The determination of the LCM need not be so lengthy as the previous illustration. When the numbers are first written as the product of *prime factors,* our work is simplified.

Recall the prime factorization of 6 and 9. $6 = 2 \cdot 3$, $9 = 3^2$. The multiples of 6: 6, 12, 18, 24, . . . could be written: $(2 \cdot 3)$, $2(2 \cdot 3)$, $3(2 \cdot 3)$, $4(2 \cdot 3)$ . . . .

Notice $(2 \cdot 3)$ is a factor of each of the multiples of 6.

The multiples of 9: 9, 18, 27, 36, 45, . . . could be written: $3 \cdot 3$, $2(3 \cdot 3)$, $3(3 \cdot 3)$, $4(3 \cdot 3)$, etc., and we see that $(3 \cdot 3)$ is a factor of each of the multiples of 9.

The common multiples of 6 and 9 are 18, 36, 54, 72, . . . and could be written $(2 \cdot 3 \cdot 3)$, $2(2 \cdot 3 \cdot 3)$, $3(2 \cdot 3 \cdot 3)$, etc. Notice that all of the foregoing common multiples of 6 and 9 contain as prime factors the prime factorization of 6, (that is, $2 \cdot 3$) and the prime factorization of 9 (that is, $3 \cdot 3$), which is $(2 \cdot 3 \cdot 3)$, but the least common multiple contains no extra factors. It contains the factors of 6, which are underlined here: $(\underline{2 \cdot 3} \cdot 3)$ and, the factors of 9 which are underlined here: $(2 \cdot \underline{3 \cdot 3})$, and only these.

Then, the L.C.M is the smallest number whose prime factors contain those of the given numbers.

Let us use this method now to determine the LCM of 12 and 18.

First determine the prime factors of the two numbers. The prime factors of the numbers, in this case 12 and 18, are grouped as follows:

12	2,	2,	3	
18		2,	3,	3

The number 2 occurs twice as a prime factor of 12 and only once as a prime factor of 18, so the LCM of 12 and 18 will contain $2 \cdot 2$. Also notice that the number 3 is found as a factor twice in 18 but only once in 12, indicating that $3 \cdot 3$ will be factors in the required L.C.M.

Thus, the LCM of 12 and 18 is $\underbrace{2 \cdot 2 \cdot 3}_{12}$ with $2 \cdot \overbrace{2 \cdot 3 \cdot 3}^{18}$, that is, $2 \cdot 2 \cdot 3 \cdot 3$, which is 36.

The next example has a L.C.M containing some factors found in only one of the numbers.

Consider the numbers 420 and 198.

The prime factors of 420 are 2, 2, 3, 5, and 7, and the prime factors of 198 are 2, 3, 3, and 11.

In table form these are written:

Number	Prime factors				
420	2, 2,	3,	5,	7	
198	2,	3,3,			11

The L.C.M is $2 \cdot 2 \cdot 3 \cdot 3 \cdot 5 \cdot 7 \cdot 11$. Notice that we have used each factor the greatest number of times it has occurred in either number.

Now let us find the L.C.M of 18 and 21.

Number	Prime factors		
18	2,	3, 3	
21		3,	7

The L.C.M is $2 \cdot 3 \cdot 3 \cdot 7$, which is 126.

This method works equally well with 3 numbers.

Number	Prime factors		
14	2,	7	
98	2,	7, 7	
44	2, 2		11

The prime factors of the LCM are 2, 2, 7, 7, and 11.

The L.C.M is 2156.

EXERCISES

Find the LCM of these sets of numbers.

	Number	Prime Factors	LCM (in factored form)	LCM
1.	20			
	14			
2.	44	2, 2, 11		
	6	2, 3	$2 \cdot 2 \cdot 3 \cdot 11$	132
3.	50			
	16			
4.	6			
	8			
5.	3			
	4			
6.	6			
	4			
7.	6			
	3			
8.	3			
	8			
9.	2			
	4			
10.	2			
	8			

Chapter 11

ADDITION AND SUBTRACTION OF FRACTIONS

Earlier chapters have discussed the operations of multiplication and division of fractions. Also we have learned to simplify fractions and change mixed numbers to improper fractions and improper fractions to mixed numbers. Now we will discuss the operations of addition and subtraction of fractions.

In Chapter 6 we added and subtracted like fractions. The following exercises will help you review this topic.

EXERCISES

Perform the indicated operations.

1. $\frac{2}{3} + \frac{5}{3} - \frac{1}{3} =$ ______

2. $\frac{5}{8} - \frac{3}{8} + \frac{7}{8} =$ ______

3. $1\frac{4}{5} + \frac{3}{5} + \frac{2}{5} =$ ______

4. $\frac{7}{16} - \frac{3}{16} + \frac{5}{16} =$ ______

5. $1\frac{5}{12} + \frac{1}{12} - \frac{6}{12} =$ ______

6. $5\frac{9}{10} + \frac{7}{10} + \frac{3}{10} =$ ______

Since we know how to add like fractions, we will now concern ourselves with addition of unlike fractions.

Let's consider the problem $\frac{7}{8} + \frac{5}{12}$.

If we recall the meaning of a fraction we recognize that this problem is asking for the sum of 7 parts out of 8 and 5 parts out of 12.

A little thought will convince us of an extremely important fact: $\frac{7}{8}$ and $\frac{5}{12}$ *cannot be combined in their present form.* The fact is that *no two fractions can be added unless they are like fractions.*

This statement is the key to addition of fractions. If only like fractions can be added then our task in combining $\frac{7}{8}$ and $\frac{5}{12}$ is to rename them so that they will be like fractions.

What are some other names for $\frac{7}{8}$?

$$\frac{7}{8} = \frac{}{16} = \frac{}{24} = \underline{\qquad} = \underline{\qquad}$$

What are some other names for $\frac{5}{12}$?

$$\frac{5}{12} = \frac{}{24} = \frac{}{36} = \underline{\qquad} = \underline{\qquad}$$

From the lists of names for $\frac{7}{8}$ and $\frac{5}{12}$, choose names that are like fractions.

$\frac{7}{8} = \underline{\qquad}$ $\qquad$ $\frac{5}{12} = \underline{\qquad}$.

We can now add $\frac{21}{24}$ and $\frac{10}{24}$, for they are like fractions.

$$\frac{21}{24} + \frac{10}{24} = \frac{31}{24} = 1\frac{7}{24}$$

So $\frac{7}{8} + \frac{5}{12} = 1\frac{7}{24}$

We now need to find the fastest way to rename two or more fractions so that they will be like fractions.

In the previous chapter we studied multiples and the least common multiple of a given set of numbers. If we rename a fraction, by multiplying the numerator and denominator by the same number, all the denominators will be multiples of the original denominator.

Consider the fractions $\frac{2}{5}$ and $\frac{4}{9}$.

What are some multiples of 5?

5, ____, ____, ____, ____, ____, ____, ____, ____

Of 9?

9, ____, ____, ____, ____, ____

What number is a common multiple of 5 and 9? _____

Can $\frac{2}{5}$ be changed to a fraction which has 45 as a denominator?

$$\frac{2}{5} = \frac{\quad}{45}$$

Can $\frac{4}{9}$ be changed to a fraction that has 45 as a denominator?

$$\frac{4}{9} = \frac{\quad}{45}$$

$$\frac{2}{5} + \frac{4}{9} = \frac{\quad}{45} + \frac{\quad}{45} = \frac{\quad}{45}$$

Like fractions are fractions that have a common denominator the *Least Common Denominator* of two or more fractions is the *Least Common Multiple* of their denominators. We will abbreviate Least Common Denominator as L.C.D.

Examples: Find the L.C.D. of $\frac{2}{3}, \frac{3}{5}, \frac{2}{9}$.

The L.C.M. of 3, 5, and 9 is 45. So the L.C.D. of $\frac{2}{3}, \frac{3}{5}, \frac{2}{9}$ is 45.

Find the L.C.D. of $\frac{1}{2}, \frac{3}{8}, \frac{5}{12}$.

The L.C.M. of 2, 8, and 12 is 24. So the L.C.D. of $\frac{1}{2}, \frac{3}{8}, \frac{5}{12}$ is 24.

EXERCISES

Find the L.C.D. of each of the following sets of fractions.

1. $\frac{5}{8}, \frac{3}{4}$ L.C.D. = ______

5. $\frac{1}{3}, \frac{1}{4}, \frac{1}{12}$ L.C.D. = ______

2. $\frac{2}{3}, \frac{5}{9}, \frac{3}{12}$ L.C.D. = ______

6. $\frac{1}{2}, \frac{2}{8}, \frac{3}{4}$ L.C.D. = ______

3. $\frac{5}{7}, \frac{2}{3}, \frac{1}{6}$ L.C.D. = ______

7. $\frac{1}{3}, \frac{1}{8}, \frac{1}{12}$ L.C.D. = ______

4. $\frac{9}{11}, \frac{3}{7}$ L.C.D. = ______

8. $\frac{5}{9}, \frac{1}{3}, \frac{3}{5}$ L.C.D. = ______

When is the L.C.D. equal to the product of the denominator?

____

Is it necessary to change the form of fractions to their *least* common denominator to add them or will *any* common denominator work?

____

What are some reasons why we should find the L.C.D.? ______________________

____

Example: $\frac{7}{9} + \frac{3}{12}$

$$\frac{7}{9} + \frac{3}{12} = \frac{28}{36} + \frac{9}{36}$$

$$= \frac{37}{36} = 1\frac{1}{36}$$

$9 = 3^2$
$12 = 2^2 \cdot 3$
L.C.M. for 9 and 12 is
$3^2 \cdot 2^2 = 36$

A shorter method of writing this solution is possible by using the common denominator only once.

$$\frac{7}{9} + \frac{3}{12} = \frac{28 + 9}{36} = \frac{37}{36} = 1\frac{1}{36}$$

Example: $$\frac{2}{3} + \frac{4}{5} + \frac{3}{10} = \frac{20 + 24 + 9}{30} = \frac{53}{30} = 1\frac{23}{30}$$

Example: $$\frac{1}{2} + \frac{5}{8} - \frac{2}{3} = \frac{12 + 15 - 16}{24} = \frac{11}{24}$$

EXERCISES

Combine the following fractions.

1. $\frac{2}{3} + \frac{5}{8} - \frac{1}{2} =$

2. $\frac{5}{16} + \frac{7}{8} + \frac{1}{4} =$

3. $\frac{3}{32} + \frac{1}{2} - \frac{1}{4} =$

4. $\frac{1}{5} + \frac{2}{3} + \frac{1}{7} =$

5. $\frac{7}{10} + \frac{12}{3} + \frac{2}{5} =$

6. $\frac{9}{25} + \frac{3}{100} + \frac{4}{10} =$

7. $\frac{5}{16} + \frac{3}{4} - \frac{1}{3} =$

8. $\frac{7}{9} + \frac{2}{3} - \frac{1}{2} =$

9. $\frac{7}{24} + \frac{3}{8} + \frac{5}{6} =$

10. $\frac{14}{15} + \frac{9}{12} =$

11. $\frac{5}{8} + \frac{5}{6} + \frac{5}{16} =$

12. $\frac{2}{3} + \frac{2}{5} + \frac{2}{9} =$

13. $\frac{11}{16} - \frac{2}{3} + \frac{1}{8} =$

14. $\frac{1}{5} + \frac{2}{3} + \frac{7}{8} + \frac{1}{2} =$

15. $\frac{7}{16} + \frac{1}{2} + \frac{3}{8} - \frac{3}{32} =$

16. $\frac{7}{9} - \frac{1}{3} + \frac{2}{5} - \frac{1}{5} =$

17. $\frac{5}{2} - \frac{3}{4} + \frac{2}{10} + \frac{1}{3} =$

18. $\frac{5}{8} + \frac{2}{3} + \frac{1}{5} + \frac{3}{7} =$

We will now look at addition problems involving mixed numbers.

Example: $4\frac{1}{3} + 3\frac{2}{5}$

Method 1. Using improper fractions.

$$4\frac{1}{3} = \frac{13}{3} \text{ and } 3\frac{2}{5} = \frac{\quad}{5}$$

The problem now becomes:

$$\frac{13}{3} + \frac{17}{5} \qquad \text{The L.C.D. is } \_\_\_\_$$

$$\frac{13}{3} + \frac{17}{5} = \frac{65 + 51}{15} = \frac{116}{15} = 7\frac{11}{15}$$

Method 2. Using the associative commutative principle.

$$4\frac{1}{3} + 3\frac{2}{5}$$

To use this method we must first remember that $4\frac{1}{3}$ means $4 + \frac{1}{3}$ and $3\frac{2}{5}$ means $3 + \frac{2}{5}$.

The problem now becomes:

$$(4 + \frac{1}{3}) + (3 + \frac{2}{5})$$

Recall that the associative property of addition states that numbers to be added can be grouped in any way, and that the commutative property states that the order of addition does not change the answer.

So $(4 + \frac{1}{3}) + (3 + \frac{2}{5}) = (4 + 3) + (\frac{1}{3} + \frac{2}{5})$

$$= 7 + (\frac{1}{3} + \frac{2}{5}) = 7 + (\frac{5 + 6}{15}) = 7\frac{11}{15}.$$

The two methods of adding mixed numbers will, of course, give the same answer. However, Method 1 can be difficult to use if the whole numbers are large. The method to be used will be left to the judgment of the student.

Example: $\frac{3}{4} + \frac{7}{8} + 5\frac{1}{2}$

$$\frac{3}{4} + \frac{7}{8} + 5\frac{1}{2} = \frac{3}{4} + \frac{7}{8} + \frac{11}{2} = \frac{6 + 7 + 44}{8} = \frac{57}{8} = 7\frac{1}{8}$$

Example: $\frac{3}{4} + \frac{7}{8} + 35\frac{1}{2}$

$$\frac{3}{4} + \frac{7}{8} + 35\frac{1}{2} = 35 + \frac{3}{4} + \frac{7}{8} + \frac{1}{2} = 35 + \frac{6 + 7 + 4}{8}$$

$$= 35 + \frac{17}{8} = 35 + 2\frac{1}{8} = 37\frac{1}{8}$$

EXERCISES

Use either method.

1. $4\frac{1}{2} + 3\frac{7}{8} + \frac{1}{2} =$

2. $14\frac{1}{3} + 58\frac{5}{6} + \frac{5}{12} =$

3. $5\frac{3}{4} + 2\frac{7}{8} =$

4. $2\frac{1}{6} + 9\frac{7}{8} + \frac{2}{3} =$

5. $266\frac{2}{3} + 33\frac{1}{3} =$

6. $92\frac{1}{2} + 83\frac{2}{3} =$

7. $1\frac{3}{4} + 7\frac{2}{5} + 9\frac{1}{10} =$

8. $5\frac{9}{10} + \frac{7}{8} + \frac{3}{4} =$

9. $7\frac{3}{8} + \frac{2}{3} + 4\frac{1}{2} =$

10. $6\frac{9}{10} + 3\frac{1}{5} + 2\frac{1}{2} =$

SUBTRACTION OF FRACTIONS

Some problems on the previous pages involved subtraction and were of no great concern. However, one type of subtraction problem needs special attention.

Let's now consider the problem $15\frac{3}{8} - \frac{7}{8}$. If we look at this problem as $15 + \frac{3}{8} - \frac{7}{8}$ we immediately run into difficulty, since $\frac{7}{8}$ cannot be subtracted from $\frac{3}{8}$.

We must rewrite the problem so that we are able to perform the subtraction.

Remember that 15 can be named in many ways (5×3, $8 + 7$, etc.). We will rewrite 15 as $14 + 1$. Our problem is now $14 + 1 + \frac{3}{8} - \frac{7}{8}$, which is $14 + \frac{8}{8} + \frac{3}{8} - \frac{7}{8}$, or $14 + \frac{11}{8} - \frac{7}{8}$.

We are now able to perform the subtraction:

$$14 + \frac{11}{8} - \frac{7}{8} = 14 + \frac{4}{8} = 14\frac{1}{2}$$

$$\text{So } 15\frac{3}{8} - \frac{7}{8} = 14\frac{1}{2}$$

This process is commonly called "borrowing." We borrowed 1 from 15 and renamed 1 as $\frac{8}{8}$.

Example: $12\frac{3}{5} - 2\frac{2}{3}$

$$12\frac{3}{5} - 2\frac{2}{3} = 12 + \frac{3}{5} - 2 - \frac{2}{3} = 12 - 2 + \frac{3}{5} - \frac{2}{3}$$

$$= 10 + \frac{3}{5} - \frac{2}{3} = 10 + \frac{9}{15} - \frac{10}{15} = 9 + \frac{24}{15} - \frac{10}{15} = 9 + \frac{14}{15} = 9\frac{14}{15}$$

Example: $126\frac{1}{3} - 16\frac{4}{7}$

$$126\frac{1}{3} - 16\frac{4}{7} = 126 - 16 + \frac{1}{3} - \frac{4}{7} = 110 + \frac{1}{3} - \frac{4}{7}$$

$$= 110 + \frac{7}{21} - \frac{12}{21} = 109 + \frac{28}{21} - \frac{12}{21} = 109 + \frac{16}{21} = 109\frac{16}{21}$$

EXERCISES

Fill in the blanks.

1. $3\frac{3}{8} = 2\frac{\_\_}{8}$

2. $15\frac{4}{27} = 14\frac{\_\_}{27}$

3. $4\frac{1}{3} = 3\frac{\_\_}{3}$

4. $12\frac{9}{16} = 11\frac{\_\_}{16}$

5. $15\frac{2}{9} = 14\frac{\_\_}{9}$

6. $8\frac{11}{12} = 7\frac{\_\_}{12}$

EXERCISES

Perform the indicated operations.

1. $3\frac{3}{8} - 1\frac{3}{4} =$

2. $15\frac{4}{27} - 12\frac{1}{9} =$

3. $4\frac{1}{3} - 2\frac{1}{2} =$

4. $12\frac{9}{16} - 7\frac{5}{8} =$

5. $15\frac{2}{9} - 8\frac{2}{3} =$

6. $8\frac{11}{12} - 4\frac{23}{24} =$

7. $2\frac{3}{4} + 5\frac{1}{8} - 3\frac{15}{16} =$

8. $1\frac{4}{5} + 2\frac{7}{8} + \frac{3}{4} =$

9. $8\frac{1}{2} - 6\frac{2}{3} + 1\frac{5}{6} =$

10. $3\frac{7}{8} - 2\frac{4}{5} =$

11. $117\frac{6}{10} - 2\frac{4}{5} =$

12. $83\frac{1}{2} + 97\frac{1}{4} - 2\frac{3}{4} =$

13. $6\frac{7}{21} - 3\frac{3}{4} =$

14. $68\frac{3}{4} + 12\frac{2}{10} - 11\frac{3}{5} =$

EXERCISES

1. $(1\frac{4}{5} + 2\frac{7}{8}) \times 3\frac{1}{2} =$ ________

2. $(1\frac{4}{5} \times 3\frac{1}{2}) + (2\frac{7}{8} \times 3\frac{1}{2}) =$ ________

3. $(14\frac{1}{5} - 2\frac{1}{2}) \times 1\frac{1}{4} =$ ________

4. $(14\frac{1}{5} \times 1\frac{1}{4}) - (2\frac{1}{2} \times 1\frac{1}{4}) =$ ________

5. $(2\frac{1}{3} + 6\frac{1}{2} - 1\frac{3}{4}) \div \frac{4}{5} =$ ________

6. $(4\frac{1}{2} - 2\frac{1}{3} + 3\frac{1}{5}) \times 2\frac{1}{2} =$ ________

7. $(4\frac{1}{2} \times 2\frac{1}{2}) - (2\frac{1}{3} \times 2\frac{1}{2}) + (3\frac{1}{5} \times 2\frac{1}{2}) =$ ______

8. $(\frac{3}{5} - \frac{1}{4}) \div \frac{2}{3} =$ ______

9. $(\frac{3}{5} \div \frac{2}{3}) - (\frac{1}{4} \div \frac{2}{3}) =$ ______

10. Bill needed two pieces of lumber $3\frac{1}{3}$ ft long and 1 piece $4\frac{1}{2}$ ft long. How much would he have left from a piece 12 ft long?

Chapter 12

OPERATIONS WITH DECIMAL FRACTIONS AND PERCENT

We are aware that numbers such as 346 mean:

$$3 \times 100 \quad + \quad 4 \times 10 \quad + \quad 6 \times 1$$

This idea is extended to the decimal notation for numerals. Some place values are indicated below.

tenths	hundredths	thousandths

We read	.7	7 tenths
	.04	4 hundredths
	.29	29 hundredths
	.003	3 thousandths
	.135	135 thousandths

The place notation indicates that a number such as 135 could be written:

$$.135 = 1 \times \frac{1}{10} + 3 \times \frac{1}{100} + 5 \times \frac{1}{1000}$$

$$\text{or } \frac{1}{10} + \frac{3}{100} + \frac{5}{1000}$$

Similarly, .402 means:

$$4 \times \frac{1}{10} + 0 \times \frac{1}{100} + 2 \times \frac{1}{1000}$$

$$\text{or } \frac{4}{10} + \frac{0}{100} + \frac{2}{1000}$$

Notice that decimal notation is a method of writing *fractions;* hence we refer to them as *decimal fractions.* Every decimal fraction can be written as a sum of common fractions in the foregoing manner.

Write these decimals as a sum of common fractions.

1. .6 __________
2. .34 __________
3. .03 3/100
4. .007 __________
5. .143 __________
6. .0007 7/10000
7. .0014 __________
8. .0136 __________
9. .7550 __________
10. .0300 __________

Do the decimals in problems 3 and 10 indicate the same number? ______

Notice from problem 6 that the fourth position indicates the number of ten-thousandths. What would the fifth position indicate?

____

____

Write in words the meaning of the following numbers. Some answers are given.

1. .7 Seven tenths
2. .34
3. .5
4. .05
5. .014
6. .0037
7. .822
8. .202
9. .9234
10. .10237 ten thousand two hundred thirty seven hundred thousandths

All of the decimals in these problems have a value between 0 and 1. The following number line indicates the approximate position of some of these numbers.*

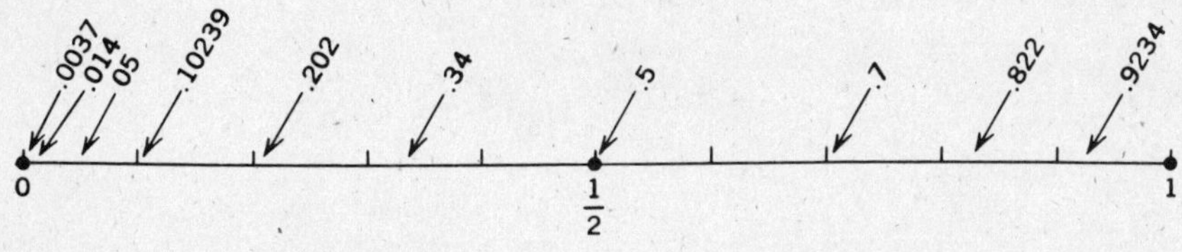

*The distance from 0 to 1 has been divided into tenths to assist in locating these points.

For greater ease in locating these points, each tenth could be divided into ________ smaller parts. These parts would then be ________ of a unit.

On the line below match the decimal with the points on the line. In the appropriate circle write the letter that precedes the decimal.

A. .750 F. .621
B. .24 G. .1
C. .99 H. .809
D. .41 I. .34
E. .5201 J. .655

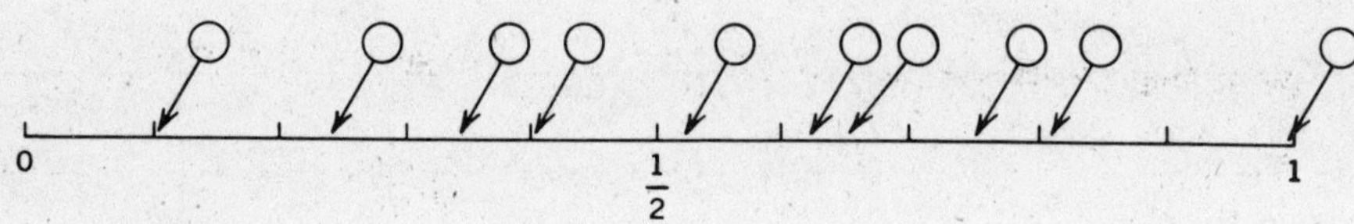

Would it be possible for two different (unequal) decimal fractions to identify a single point on the number line? ________

When the decimal fraction is preceded by a counting number, we have the decimal representation of a mixed number. Examples are 2.46, 34.2, and 1003.742.

From writing decimals earlier in this chapter and from writing out counting numbers in Chapter 1, you have noticed that the word "and" was not used. This word is reserved to indicate a *decimal point* when reciting a mixed number. For example, we read 234 as "two hundred thirty four," and .163 as "one hundred sixty three thousandth. However, 12.437 is read as "twelve *and* four hundred thirty seven thousandth, and 4.3 is read as "four *and* three tenths."

The *expanded* notation of a number such as 43.702 is:

$$4 \times 10 + 3 \times 1 + 7 \times \frac{1}{10} + 0 \times \frac{1}{100} + 2 \times \frac{1}{1000}$$

$$\text{or } 40 + 3 + \frac{7}{10} + \frac{2}{1000}$$

This idea, the expanded notation of a number, will aid our understanding of the operations on decimal fractions. The basic operations are: (a) addition, (b) subtraction, (c) multiplication, (d) division.

Addition of Decimals

(a) .3 + .5 could be written $\frac{3}{10} + \frac{5}{10}$. Therefore .3 + 5 $= \frac{8}{10}$, which can easily be written in decimal notation as .8.

(b) .12 + .35 could be written $\frac{1}{10} + \frac{2}{100} + \frac{1}{10} + \frac{5}{100}$, which equals $\frac{4}{10} + \frac{7}{100}$, which is .47

The best manner of adding these decimals is to make a vertical column of numbers, keeping the decimal points lined up one under the other. Thus:

(a)	(b)	(c)	(d)
.3	.12	.34	.67
.5	.35	.62	.31
.8	.47		

This alignment of the decimals permits us to add numbers with *common* place values, or fractions with *common* denominators.

In (a) above we are adding $\frac{3}{10}$ to $\frac{5}{10}$. This is simple addition problem because the *denominators* are the same. In (b) we add first $\frac{2}{100}$ to $\frac{5}{100}$, placing the 7 in the proper (second) position; then add $\frac{1}{10}$ and $\frac{3}{10}$ and place the 4 (the sum of the numerators of the fractions) in the first position to the right of the decimal point.

The notion of "carrying" applies to the addition of decimals just as it did when adding counting numbers.

Remember that:

$1 = \frac{10}{10}$ $\qquad \frac{1}{100} = \frac{10}{1000}$

$\frac{1}{10} = \frac{10}{100}$ $\qquad \frac{1}{1000} = \frac{10}{10{,}000}$

In the following example two decimal fractions are added on the left and these same two numbers expressed as a sum of common fractions are added at right.

Example:

$$\begin{array}{r} 1 \\ .219 \\ \underline{.347} \\ .566 \end{array} \qquad \begin{array}{l} \frac{2}{10} + \frac{1}{100} + \frac{9}{1000} \\ \underline{\frac{3}{10} + \frac{4}{100} + \frac{7}{1000}} \\ \frac{5}{10} + \frac{5}{100} + \frac{16}{1000} \end{array}$$

$$= \frac{5}{10} + \frac{5}{100} + \frac{10}{1000} + \frac{6}{1000}$$

$$= \frac{5}{10} + \frac{5}{100} + \frac{1}{100} + \frac{6}{1000}$$

$$= \frac{5}{10} + \frac{5}{100} + \frac{1}{100} + \frac{6}{1000}$$

$$= \frac{5}{10} + \frac{6}{100} + \frac{6}{1000}$$

$$= .566$$

Notice that $\frac{16}{1000}$ is changed to $\frac{10}{1000} + \frac{6}{1000}$, then the $\frac{10}{1000}$ is changed to $\frac{1}{100}$ and added to the $\frac{5}{100}$. At this stage we have carried the one (in both the decimal fraction and the common fraction).

Example:

$$\begin{array}{r} .34 \\ \underline{.82} \\ 1.16 \end{array} \qquad \begin{array}{l} \frac{3}{10} + \frac{4}{100} \\ 8 \\ \underline{\frac{8}{10} + \frac{2}{100}} \\ \frac{11}{10} + \frac{6}{100} \end{array}$$

$$= \frac{10}{10} + \frac{1}{10} + \frac{6}{100}$$

$$= 1 + \frac{1}{10} + \frac{6}{100}$$

$$= 1.16$$

Add the following decimal fractions. You should do all of these correctly in four minutes.

1. 43.62 17.34	2. 6.007 .04	3. 11.703 2.3	4. 9.98 .76
5. 103.6 .37	6. 13.42 3.42	7. 41.37 6.07	8. 3.6 2.08 9.12
9. 3.0063 .74 6.667 .14	10. 29.12 7.40 4.003	11. 1.3471 .7342 30.0463 6.0936	12. 1.8 2.86 9.3
13. 764.11 98.32	14. 9.103 .269 .80 .22	15. .723 4.6 9.302 .54	16. 6.1 2.3 5.8 3.2

Subtraction of Decimals

Subtraction of decimal fractions is very similar to subtraction of counting numbers. Decimal points are kept in a vertical line and borrowing is frequently necessary.

Let's consider the problem 2.4 − .7. The result (difference) is 1.7. The borrowing process is shown below both with decimals and where the same two numbers are first changed to common fractions.

Example:

$$\begin{array}{r} {\scriptstyle 1} \\ \not{2}.4 \\ \underline{.7} \\ 1.7 \end{array}$$

$$\begin{array}{lll} 2 + \frac{4}{10} \rightarrow & 1 + \frac{10}{10} + \frac{4}{10} \rightarrow & 1 + \frac{14}{10} \\ \underline{\quad\ \ \frac{7}{10}} \rightarrow & \underline{\qquad\qquad \frac{7}{10}} \rightarrow & \underline{\quad\ \ \frac{7}{10}} \\ & & 1 + \frac{7}{10} = 1.7 \end{array}$$

Example:

$$\begin{array}{r} {\scriptstyle 2\ 3} \\ \not{3}0.\not{4}2 \\ \underline{1.36} \\ 29.06 \end{array}$$

$$\begin{array}{ll} 30 + \frac{4}{10} + \frac{2}{100} \rightarrow & 30 + \frac{3}{10} + \frac{1}{10} + \frac{2}{100} \\ \underline{\ \ 1 + \frac{3}{10} + \frac{6}{100}} \rightarrow & \underline{\ \ 1 + \frac{3}{10} + \frac{6}{100}\qquad} \end{array}$$

$$\begin{array}{ll} \rightarrow 30 + \frac{3}{10} + \frac{12}{100} \rightarrow & 29 + 1 + \frac{3}{10} + \frac{12}{100} \\ \rightarrow \underline{\ \ 1 + \frac{3}{10} + \frac{6}{100}} \rightarrow & \underline{\qquad\ 1 + \frac{3}{10} + \frac{6}{100}} \\ & \qquad 29 + \frac{6}{100} = 29.06 \end{array}$$

EXERCISES

Subtract

1. 42.66 − .40
2. 3.12 − .03
3. 3.14159 − .624
4. .6439 − .1027
5. 38.214 − 17.303
6. 123.00 − .04
7. 14.394 − 2.93
8. 142.640 − 37.730
9. 14.222 − .364
10. $4.16 − .13
11. $14.76 − 3.95
12. $107.20 − 95.60
13. $12.36 − 2.50
14. $88.43 − 9.99
15. 7220.00 − 46.31

We have been writing decimal fractions as a *sum* of common fractions with denominators of 10, 100, 1000, etc.

When reciting a decimal fraction or writing it in words, we are reminded that the decimal could be expressed as a single common fraction. For instance, .73 could be written $\frac{73}{100}$; .123 could be written $\frac{123}{1000}$.

Fill in the blank spaces below by writing each decimal numeral as a common fraction or mixed number.

1. .7 = ____________________
2. .5 = ____________________
3. .34 = ____________________
4. .19 = ____________________
5. .78 = ____________________
6. .03 = ____________________
7. .05 = ____________________
8. .106 = ____________________
9. .236 = ____________________
10. .100 = ____________________
11. .005 = ____________________
12. .1036 = ____________________
13. .2464 = ____________________
14. .0001 = ____________________
15. 1.4 = ____________________
16. 1.06 = ____________________
17. 1.003 = ____________________
18. 2.67 = ____________________

Write these as decimal fractions.

1. $\frac{6}{10}$ = .6 ____________
2. $\frac{13}{100}$ = ____________
3. $\frac{5}{10}$ = ____________
4. $\frac{5}{100}$ = ____________
5. $\frac{5}{1000}$ = ____________
6. $\frac{14}{100}$ = ____________
7. $\frac{163}{1000}$ = ____________
8. $\frac{429}{1000}$ = ____________
9. $\frac{2666}{10,000}$ = ____________
10. $\frac{2003}{10000}$ = ____________
11. $\frac{7}{10}$ = ____________
12. $\frac{7}{100}$ = ____________
13. $\frac{7}{1000}$ = ____________
14. $\frac{7}{10000}$ = ____________
15. $\frac{14}{10,000}$ = ____________
16. $\frac{20}{10}$ = ____________
17. $\frac{24}{10}$ = ____________
18. $\frac{119}{100}$ = ____________
19. $\frac{2146}{1000}$ = ____________
20. $\frac{36009}{10,000}$ = ____________

Powers of ten occur frequently in mathematics and science. In fact each of the denominators above is a power of ten. 10^2 is the second power of ten; 10^3 is the third power of ten, and so forth. The small numeral is called the exponent.

You recall from an earlier chapter that:
$10^1 = 10$; $10^2 = 100$; $10^3 = 1000$; $10^4 = 10000$; $10^5 = 100000$.

Using this pattern it appears that the number of zeros in the product is equal to the ____ ________ of ten.

In problem 20 above your answer should be 3.6009. The decimal point was moved four places to the left. There are four zeros in the denominator. Does this occur in all problems?

Multiplication of Decimals

Multiplication of decimal fractions is done by first ignoring the decimal points and proceeding as if we were simply multiplying counting numbers. Next, to locate the decimal point in the product, we find the sum of the decimal places in the two factors and count off this number of places in the product. Count from right to left.

The reason for locating the decimal point in this fashion is seen when multiplication of two decimal fractions is compared to the multiplication of the same two numbers as common fractions.

Example:

$$\begin{array}{r} .43 \\ \underline{.62} \\ 86 \\ \underline{258} \\ .2666 \end{array} \qquad \frac{43}{100} \times \frac{62}{100} = \frac{2666}{10{,}000} = .2666$$

Example:

$$\begin{array}{r} 1.47 \\ \underline{.9} \\ 1.323 \end{array} \qquad \frac{147}{100} \times \frac{9}{10} = \frac{1323}{1000} = 1.323$$

Example:

$$\begin{array}{r} .0073 \\ \underline{1.7} \\ 511 \\ \underline{73} \\ .01241 \end{array} \qquad \frac{73}{10{,}000} \times \frac{17}{10} = \frac{1241}{100{,}000} = .01241$$

Multiply:

1. $\frac{1}{2} \times \frac{1}{3} =$ __________

2. $\frac{3}{7} \times \frac{4}{3} =$ __________

3. $\frac{7}{10} \times \frac{3}{10} =$ __________

4. $\frac{14}{100} \times \frac{16}{3} =$ __________

5. $3\frac{1}{2} \times 2\frac{1}{4} =$ __________

6. $\begin{array}{r} .21 \\ \underline{.47} \end{array}$

7. $\begin{array}{r} 2.36 \\ \underline{1.9} \end{array}$

8. $\begin{array}{r} 146 \\ \underline{2.9} \end{array}$

9. $\begin{array}{r} .007 \\ \underline{.03} \end{array}$

10. $\begin{array}{r} 14.29 \\ \underline{30.9} \end{array}$

11. $\begin{array}{r} 77.63 \\ \underline{4.392} \end{array}$

Division of Decimal Fractions

If 37.5 were divided into 5 equal parts, how large would each part be?

This question might be answered by (a) changing 37.5 to $37\frac{1}{2}$ and dividing according to methods previously established, or (b) dividing the decimal fraction 37.5 directly by 5. Both processes are shown here.

(a) $37\frac{1}{2} \div 5$

is $\frac{75}{2} \div 5$ which is $\frac{\overset{15}{\cancel{75}}}{2} \cdot \frac{1}{\cancel{5}} = \frac{15}{2} = 7\frac{1}{2}$

(b)
$$5\overline{)37.5}^{\;7.5}$$

Both answers are, of course, the same. Division of decimal fractions by counting numbers always follows the above plan. The decimal point is placed directly above its position in the dividend.

A word must be said here regarding neatness of your work. Neatness is always important, but in division it is a necessity. Unless you keep the numbers in the quotient directly over the numbers in the dividend, you may place the decimal point incorrectly.

The quotient (the result of division) will always yield a precise decimal fraction when the prime factors of the divisor are 2's or 5's (or both). For example, if we divide any decimal fraction by 2, 4, 5, 8, 10, 16, 20, 25, 50, 125, etc., we can be assured of an exact quotient since the prime factorization of all of these numbers contains only 2's and 5's. For instance $20 = 2^2 \cdot 5$.

The following division problems illustrate this.

1.
```
       .26
  25 | 6.50
       5 0
       1 50
       1 50
```

2.
```
      .4275
  8 | 3.4200
      3 2
        22
        16
         60
         56
          40
          40
```

EXERCISES

Complete the following: (zeros may be added to the decimal fraction without changing the number. This will be necessary in problem 1 and others below.)

1. $16\overline{)4.20}$

2. $25\overline{)75.25}$

3. $40\overline{)136.0}$

4. $125\overline{)1000.0}$

5. $64\overline{)2.80}$

6. $20\overline{)63.6}$

When no decimal point is written it is assumed to be to the right of a counting number. All of these are different forms of the same number:

17 17. 17.0 17.00 17.000

If the divisor contains any prime factor other than 2 and 5, which is not a factor of the number being divided, the quotient will have infinitely many decimal places. In practice we stop dividing when we obtain the degree of precision needed.

$$\text{Consider} \quad 3\overline{\big)\,10.40}\quad \begin{matrix}3.466666\cdots\end{matrix}$$

Each time another zero is added in the dividend another 6 occurs in the quotient. This process does not end. This is called a "non-terminating decimal."

We may wish to "round off" this number to thousandths or to the "nearest thousandth." We do this by considering first the number 3.4666. This number is between 3.4660 and 3.4670, but closer to the second number, 3.4670. Thus we "round off" 3.4666 to 3.467.

If we wished to round off 1.42674 to thousandth (3 places) we first write ____________ (4 places). This last number is between *1.4260* and ____________. Therefore ____________ rounded off would be approximately equal to ____________.

We could write this 1.42674 ≈ 1.427. The wiggly lines mean "approximately equal to."

Should the last digit be 5 or "half way between," it is customary to round off to the even number.

Thus 14.245 rounded off to *hundredth* would be 14.24, 4 being the even number but 14.255 would be 14.26, since 6 is even.

4.683 to the nearest hundredth is 4.68
1.585 to the nearest hundredth is 1.58
7.618 to the nearest tenth is 7.6
.5834 to the nearest thousandth is .583

EXERCISES

Round off these numbers to the nearest *hundredth.*

1. 2.759 2.76
2. .314 .31
3. .888 .89
4. .634 .63
5. .996 1.00
6. 8.316 8.32
7. 2.465 2.46
8. 19.835 19.83
9. 7.0486 7.05
10. 3.009 3.01

Round off to the nearest *tenth.*

1. 4.22 4.2
2. 1.79 1.8
3. 13.401 13.4
4. 10.006 10.0
5. 3.42 3.4
6. .65 .6
7. .83 .8
8. .754 .7
9. 3.14159 3.14
10. 3.001 3.0

Circle the correct statement.

a. $3.62 = 3.6$

b. $3.62 \approx 3.6$ (circled)

EXERCISES

Perform the indicated division. Stop dividing when four decimal places have been attained, then round off the quotient correct to thousandth.

1. $7\overline{)1.2347}$ Answer ____________

2. $3\overline{)12.7346}$ Answer ____________

3. $14\overline{)8.99346}$ Answer ____________

4. $23\overline{)49.39618}$ Answer ____________

5. $74\overline{)62.14224}$ Answer ____________

6. $3\overline{)100}$ Answer ____________

7. $18\overline{)0.0414}$ Answer ____________

8. $564\overline{)2.1432}$ Answer ____________

Divide and round off to *hundredths.*

9. $203\overline{)303688}$ Answer ____________

10. $14\overline{)8.862}$ Answer ______

11. $48\overline{)614.208}$ Answer ______

12. $60\overline{)808.920}$ Answer ______

All the division we have been dealing with so far has involved a *divisor* that is a *counting* number. There is little trouble, however, in working with divisors that are decimal fractions. Recall that a common fraction remains unchanged when we multiply both numerator and denominator by any non-zero counting number.

$\frac{7}{3} = \frac{21}{9}$ We have multiplied numerator and denominator by 3.

$\frac{20}{33} = \frac{40}{66}$ Multiplying the numerator and denominator by 2.

Now consider the problem of dividing 6.30 by .4.

This could be shown:

$$.4\overline{)6.30} \quad \text{or} \quad \frac{6.30}{.4}$$

The second number above can easily be changed to an equivalent number by multiplying *both* numerator and denominator by 10. Ten is selected as a multiplier because the product of 10 and .4 is 4, a *counting* number.

Therefore $\frac{6.30}{.4} = \frac{63.0}{4}$, or $4\overline{)63.0}$.

We are familiar with the method for dividing this.

Divide 4.364 by .21:

$$\frac{4.364}{.21}$$

Here we multiply numerator and denominator by 100 since this will make the ________ a counting number.

Our division problem then will look like this:

$$21\overline{)436.4}$$

In the following quotients indicate the number by which you would multiply both numerator and denominator to make the denominator a counting number.

1. $\dfrac{17}{30.4}$ Answer 10

2. $\dfrac{80.46}{.27}$ 100

3. $\dfrac{21}{.003}$ 1000

4. $\dfrac{863.9}{.127}$ 1000

5. $\dfrac{16309}{2.2}$ ______

6. $\dfrac{30.997}{.4362}$ ______

7. $\dfrac{12.894}{.2}$ 10

8. $\dfrac{31496}{.297}$ ______

9. $\dfrac{248.3}{4.63}$ ______

10. $\dfrac{8.9}{2.007}$ ______

It is not necessary, however, to write a division problem in this fractional form before dividing. By observing the position of the decimal point in the divisor, we can determine whether we must multiply divisor and dividend by 10, 100, 1000, etc.

In the problem, $6.37\overline{)149.342}$, we multiply both divisor and dividend by ________. This gives us $637\overline{)14934.2}$.

A general rule to follow is: Move the decimal point in the divisor the number of places necessary to make it a whole number; *then* move the decimal point in the dividend exactly this same number of places.

In the following problem the decimal point is moved two places to the right in both divisor and dividend. This movement is indicated by the curved line and the *new* position of the decimal point in the dividend is indicated by a symbol $\wedge$.

$$2.67_\wedge\overline{)1.43_\wedge69}$$

It may be necessary to add zeros in the dividend in order to move the decimal point the required number of places. A problem of this type is $2.4\,/\overline{75}$, and should be written:

$$\begin{array}{r} 31.25 \\ 2.4_\wedge\overline{)75.0_\wedge00} \\ \underline{72} \\ 30 \\ \underline{24} \\ 60 \\ \underline{48} \\ 120 \\ \underline{120} \end{array}$$

EXERCISES

Divide.

1. $27.4\overline{)35.620}$ Answer ______________

2. $.25\overline{)36.50}$ Answer ______________

3. $7.5\overline{)100}$ Answer ______________

4. $17.9\overline{)41.886}$ Answer ________________

5. $.003\overline{)24.747}$ Answer ________________

Problems 6 – 10 will have nonterminating decimals. Round off answers to *hundredths.*

6. $.7\overline{)2.4298}$ Answer ________________

7. $.7\overline{)2.4298}$ Answer ____________

8. $1.2\overline{)49.0}$ Answer ____________

9. $.042\overline{).05755}$ Answer ____________

One particular disadvantage of decimal fractions is that many common fractions, such as $\frac{1}{3}$, cannot be expressed exactly as a decimal. We are restricted to an approximation such as .333. We write

$$\frac{1}{3} \approx .333$$

$$\frac{2}{3} \approx .667$$

$$\frac{1}{6} \approx .167$$

These have been rounded off to thousandths, but, as you know, could be expressed to as many decimal places as you wish.

All common fractions then, can be approximated by a decimal form. We *divide* the numerator by the denominator to determine the decimal fraction equivalent or approximation.

In the blanks below fill in the decimal equivalents (or approximations) to the given fractions.

$\frac{1}{2} =$ ______ $\frac{2}{3} \approx$ ______

$\frac{1}{3} \approx$ ______ $\frac{3}{4} =$ ______

$\frac{1}{4} =$ ______ $\frac{2}{5} =$ ______

$\frac{1}{5} =$ ______ $\frac{4}{5} =$ ______

$\frac{3}{5} =$ ______ $\frac{5}{6} \approx$ ______

$\frac{1}{6} \approx$ ______ $\frac{3}{8} =$ ______

$\frac{1}{8} =$ ______ $\frac{7}{8} =$ ______

$\frac{5}{8} =$ ______ $\frac{3}{20} =$ ______

How would you divide $\dfrac{3.47}{\frac{3}{4}}$?

There are two good ways. Either change 3.47 to $3\frac{47}{100}$ and divide by $\frac{3}{4}$ using the rules of division for common fractions; or else divide 3.47 by .75, a decimal equivalent that you have already calculated. The two problems are done below.

(a) $3\frac{47}{100} \div \frac{3}{4} = \frac{347}{100} \div \frac{3}{4} = \frac{347}{\cancel{100}_{25}} \cdot \frac{\cancel{4}^{1}}{3} = \frac{347}{75} = 4\frac{47}{75}$

(b)

```
           4.626̄
.75∧ ) 3.47∧000
       3 0
       ----
        470
        450
        ---
         200
         150
         ---
          500
          450
          ---
           50 –
```

The solution, $4\frac{47}{75}$, is exact. The solution above, $4.62\overline{6}$ is an approximation. Rounded off to hundredths we would have 4.63. The line above the 6 in (b) indicates that the 6's will repeat without end if division is continued.

EXERCISES

Divide the following. Solve by using (a) common fractions and then (b) decimals, rounded off to hundredths when necessary.

1. $\dfrac{4.2}{\frac{3}{8}}$

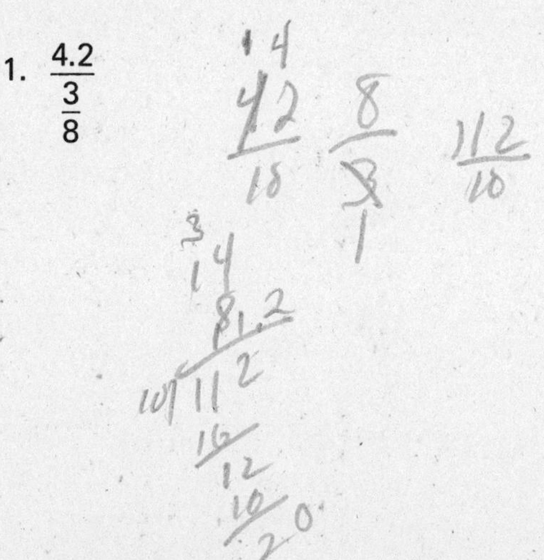

a ______ b ______

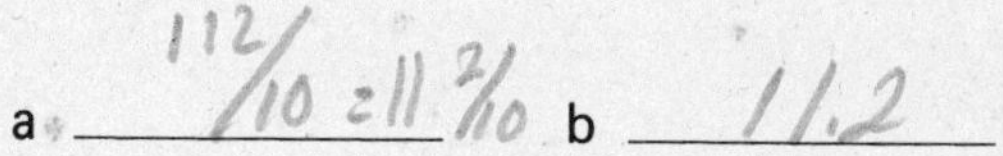

2. $\dfrac{1.63}{\dfrac{2}{3}}$ a ______ b ______

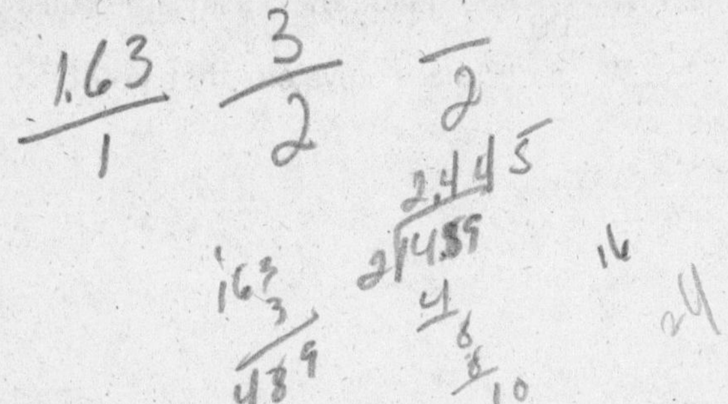

3. $\dfrac{29.7}{1\frac{4}{5}}$ a ______ b ______

4. $\dfrac{\dfrac{1}{4}}{.004}$ a ______ b ______

5. $\dfrac{\frac{3}{5}}{.14}$ a ____________ b ____________

6. $\dfrac{1\frac{3}{8}}{.6}$ a ____________ b ____________

7. $\dfrac{2.85}{1\frac{3}{5}}$ a ____________ b ____________

OPERATIONS WITH PERCENT

Percent means *hundredths.* The symbol % is read percent and means hundredths.

14% (fourteen percent) = $\frac{14}{100}$ = .14

25% (twenty five percent) = $\frac{25}{100}$ = .25

6% (six percent) = $\frac{6}{100}$ = .06

250% (two hundred fifty percent) = $\frac{250}{100}$ = 2.50

Hundredths, which is two decimal places, can always be used for the symbol %. We must agree that if a numeral has no decimal point, the decimal is understood to be after the right-most digit. The following all name the same numbers.

17 = 17. = 17.0 = 17.00 = 17.0000 etc.

Move the decimal point two places to the left.

(a) 14.5 __________ (b) 176 __________

(c) .843 __________ (d) 2.8 __________

Move the decimal point two places to the right.

(a) .35 __________ (b) 2.3 __________

(c) 84 __________ (d) 9.16 __________

Because we have already studied operations with decimal fractions our work with percent will be much easier. Our first concern is changing percents to decimals. Since percent means hundredths, the rule for this change is a simple one.

"To change a percent to an equivalent decimal, move the decimal two places to the left and remove the % sign."

Example: 54% = .54; 87.5% = .875

EXERCISES

Change these % to decimals.

1. 36% = ______		2. 54% = ______	
2. 94% = ______		9. 20% = ______	
3. 256% = ______		10. 25% = ______	
4. 4% = ______		11. 50% = ______	
5. 37.5% = ______		12. 1% = ______	
6. 2% = ______		13. 2.6% = ______	
7. 116% = ______		14. 26% = ______	

Changing from a decimal to a % will of course be the opposite of changing from a percent to a decimal.

If 26% = .26 then .84 should equal 84%.

State a rule for changing from decimal numerals to percent.

____

____

Examples: .04 = 4%; 2.61 = 261%; .035 = 3.5%.

EXERCISES

Change these decimals to %.

1. .25 = ______ %	8. .625 = ______ %
2. 2.5 = ______ %	9. .0625 = ______ %
3. .75 = ______ %	10. .01 = ______ %
4. .375 = ______ %	11. .35 = ______ %
5. 2.9 = ______ %	12. 3.5 = ______ %
6. .003 = ______ %	13. 35 = ______ %
7. .41 = ______ %	14. 1 = ______ %

Percents less than 1% can be of special concern. These really should cause no trouble but we sometimes have difficulty following the rules.

Let's state the meaning again for emphasis. Percent means ________________, which is ________ decimal places. So to change from decimals to percents or percents to decimals we always move the decimal point exactly ________ places.

EXERCISES

Fill the blanks.

1.	.005 = .5 %	8.	.15% = .0015
2.	.003 = .3 %	9.	.0025 = .25 %
3.	.0005 = .05 %	10.	.25% = .0025
4.	.5% = .005	11.	.033% = .00033
5.	.05% = .0005	12.	.0025% = .000025
6.	2.3% = .023	13.	.25 = 25 %
7.	.023% = 23	14.	.0075 = .75 %

Now that we have mastered the ability to change from decimals to percents and percents to decimals we will concern ourselves with operations with percents.

One basic rule will always be followed and it is a simple one. It merely states that we will do all of our work with % by first changing to decimals.

Example: Find 32% of 84.

32% = .32, so our problem is

$$.32 \times 84 \quad \text{or} \quad \begin{array}{r} 84 \\ \underline{.32} \\ 168 \\ \underline{252} \\ 26.88 \end{array}$$

So 32% of 84 is 26.88.

EXERCISES

1. 26% of 812 = ________

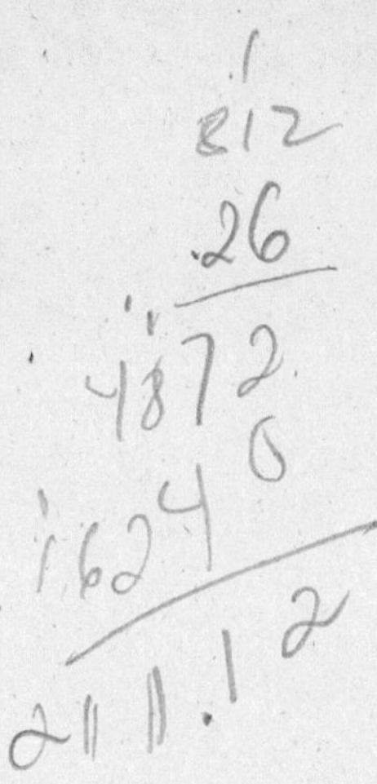

2. 5% of 9110 = ________

3. 115% of 96 = ________

4. .005% of 212 = ______________

5. $87\frac{1}{2}$% of 224 = ______________

6. .0025% of 400 = ______________

Earlier in this chapter we discussed changing common fractions to decimal equivalents. Of course, since we can now change decimal fractions to %, these two things can be combined to change from common fractions to percent.

Example Change $\frac{3}{8}$ to %.

$$\begin{array}{r} .375 \\ 8\overline{)3.000} \\ \underline{2\,4} \\ 60 \\ \underline{56} \\ 40 \\ \underline{40} \end{array}$$

$\frac{3}{8} = .375$

but $.375 = 37.5\%$

so $\frac{3}{8} = 37.5\%$

EXERCISES

Change these common fractions to %. (Round to 3 places if necessary.)

1. $\frac{3}{4}$ = ________ %

2. $\frac{7}{8}$ = ________ %

3. $\frac{6}{7}$ = ________ %

4. $\frac{5}{9}$ = ________ %

EXERCISES

1. Find $\frac{1}{2}$% of 1264.

2. Find 50% of 1264.

3. Find $\frac{3}{4}$% of 875.

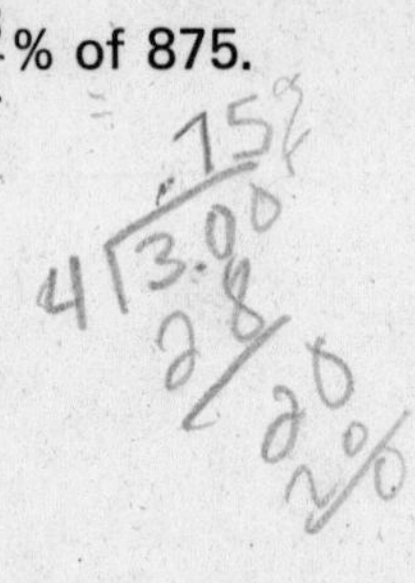

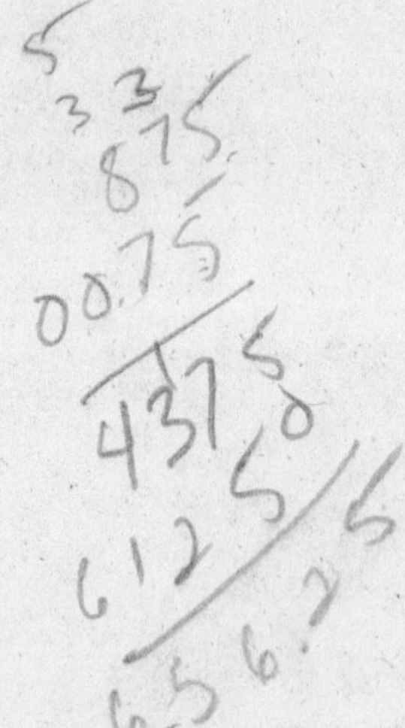

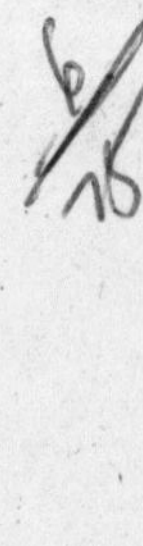

4. Find 75% of 875.

5. Find 10% of 961.

6. Find .10% of 961.

7. What % of 84 is 21? (Hint: this is asking 21 is what part of 84? or $\frac{21}{84}$ = _____?)

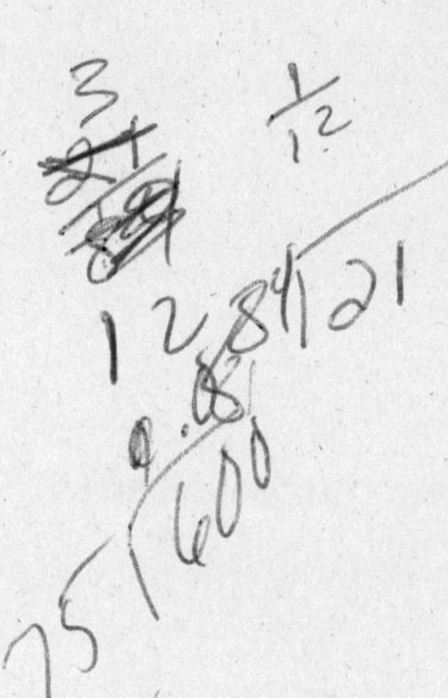

8. 15 is _____% of 60?

9. 120 is what % of 25?

10. 8 is 10% of ____? (or $.10x = 8$)

11. A store has a sale on men's wear. Everything is 20% off. If a suit was priced at $48.50, what is the sale price?

12. A finance company charges $\frac{3}{4}\%$ interest per month. What would be the amount of interest for a month on $860.00?

EXERCISES

1. Complete this table.

	Common fraction	Decimal fraction (Correct to 3 places)	Percent
(a)	$\frac{1}{2}$	______	______
(b)	$\frac{1}{3}$	______	______
(c)	$\frac{1}{4}$	______	______
(d)	$\frac{1}{5}$	______	______
(e)	$\frac{1}{6}$	______	______
(f)	$\frac{1}{7}$	______	______
(g)	$\frac{1}{8}$	______	______
(h)	$\frac{1}{9}$	______	______
(i)	$\frac{1}{10}$	______	______
(j)	______	.80	______
(k)	______	______	45%
(l)	______	.234	______
(m)	$\frac{3}{2}$	______	______
(n)	$\frac{5}{4}$	______	______
(o)	______	______	210%

2. Mary says she would like to reduce her weight by 12%. If she now weighs 138 pounds, how much does she wish to weigh?

3. Bill Jones is a dry goods salesman. He is paid a salary of $60 plus 2% of all sales over $400 each week. Last week Bill's sales totaled $968. What was his salary?

4. A grocer figures he must sell each article for 18% more than his cost in order to make a fair profit and take care of overhead expenses. What would be the selling price of an article that cost the grocer 42 cents?

5. If 18% of the students in a class made a grade of A on a certain test, how many students were in the class if there were 9 A's?

Chapter 13

OPERATIONS ON DIRECTED NUMBERS

In previous chapters we have been concerned with operations on a set of numbers that are sometimes called "the numbers of arithmetic." As we have seen before, these numbers can be represented on a diagram called a number line.

0 $\frac{1}{2}$ 1 $1\frac{2}{3}$ 2 3 4 5 6 7

Such a line is constructed with certain agreements in mind. For instance, each unit (from 0 to 1 or 5 to 6, etc.) should be the same length. Another agreement is that the larger of two numbers is to the right of the other.

Place points on this number line to represent the following:

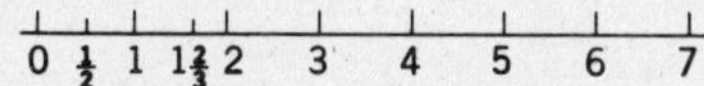

(a) 3

(b) $2\frac{1}{2}$

(c) $5\frac{1}{4}$

(d) 7

(e) $3\frac{3}{4}$

(f) $2\frac{7}{8}$

Does every point on the line have a number to represent it?

In geometry (the study of points, lines, and so on) a line is described as having no end points. In other words it has no beginning and no end. What about the points to the left of zero? Do they have numbers to represent them?

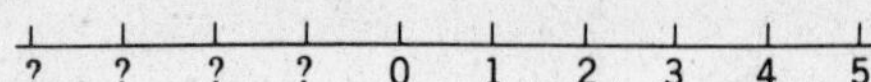

It seems logical that these points should be represented by numbers. But if such numbers exist, they will be less than zero because we have agreed that a number to the right of another number is the larger one.

The answer is that there is a set of numbers to the left of zero. They are called the negative numbers and are represented by a minus sign before the numbers, for example, -3, -4, $-2\frac{1}{2}$.

-3 should be read "negative three" or "negative of three."

−1 0 1

Place points to represent these numbers on the number line above.

(a) 4

(b) -4

(c) $2\frac{1}{2}$

(d) $-2\frac{1}{2}$

(e) -7

(f) $-1\frac{1}{2}$

(g) $\frac{1}{2}$

(h) 5

The set of numbers that can be represented on a number line is called "the set of real numbers." Therefore the line we have been using is called "the real number line."

Our task in this chapter is to learn to perform the four basic operations – addition, subtraction, multiplication, and division – on the set of real numbers.

Another name commonly used for these numbers is "signed numbers." This name arises from the fact that a sign (+ or –) precedes each number. The numbers to the right of zero are called the positive numbers and those to the left of zero are called the ________________ numbers. Still another name for this set of numbers is "the directed numbers." This name arises from the fact that a number indicates a direction.

ADDING DIRECTED NUMBERS

What does addition mean to you?

____

____

You may have listed several things. However, we wish to think of addition of directed numbers as in the following illustrations:

+4 added to +5 will mean four units to the right followed by five units to the right. So +4 added to +5 equals +9.

+4 added to −5 will mean 4 units to the ________ followed by 5 units to the ________. The end result leaves us at 1 unit to the ________. So +4 added to −5 equals −1.

We can use a number line to represent these sums.

+4 added to +5 = +9

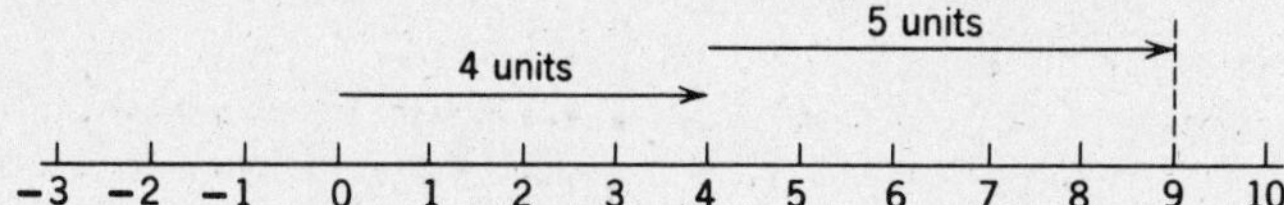

+4 added to −5 = −1

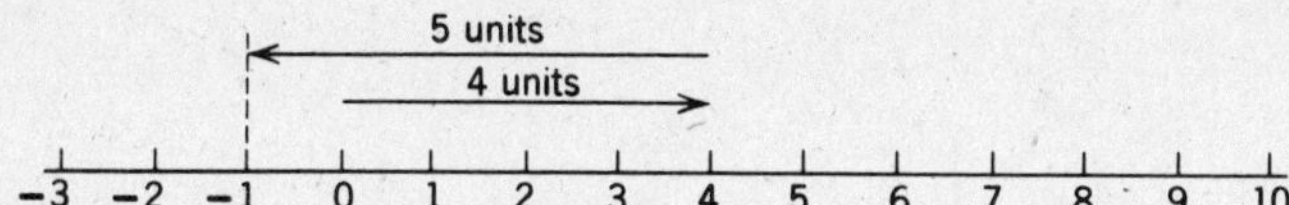

EXERCISES

Draw arrows to show the following sums.

1. +2 added to +4

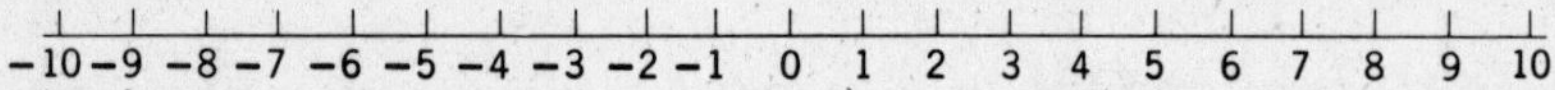

2. +6 added to −3

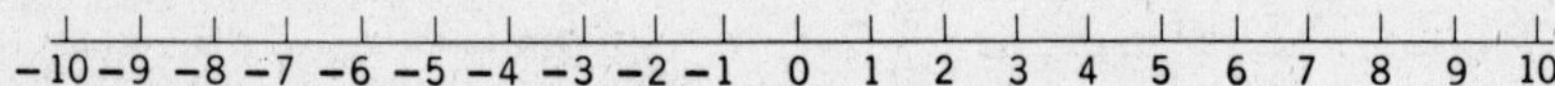

3. +3 added to −6

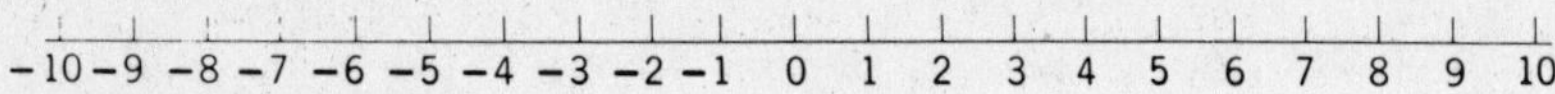

4. −4 added to −2

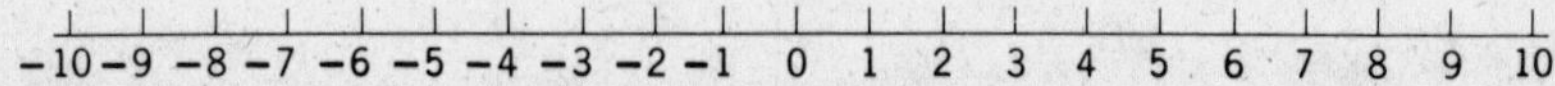

5. +5 added to +3

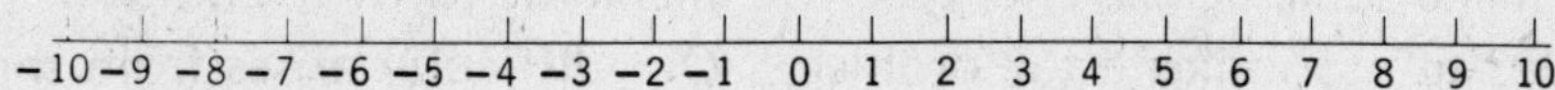

6. +7 added to −12

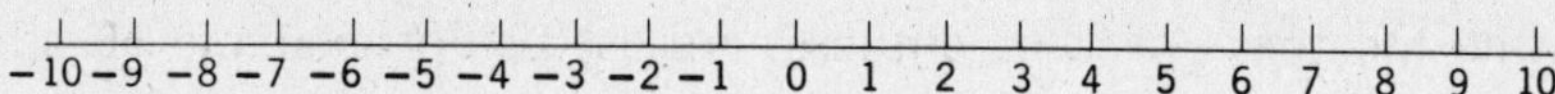

7. +8 added to −6

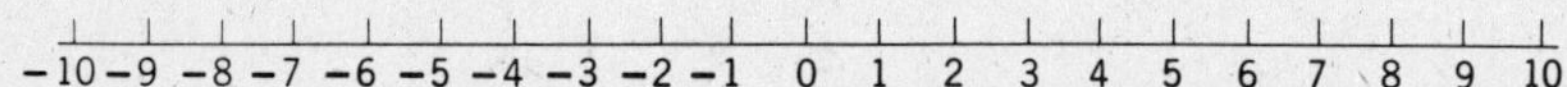

8. −3 added to −7

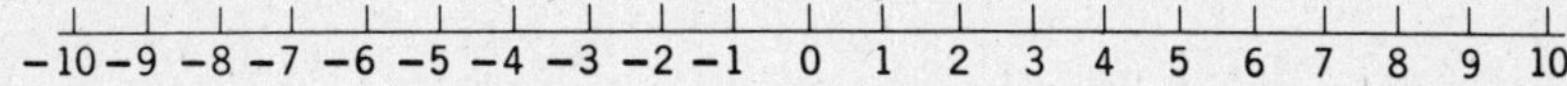

9. −8 added to +14

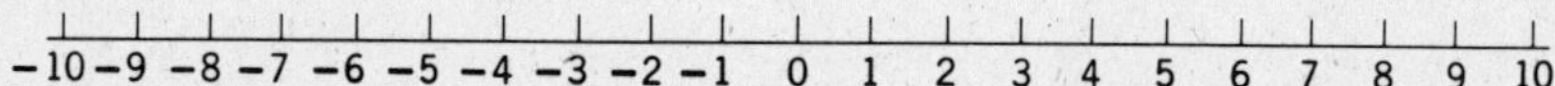

If we take a look at your results from the previous problems we can establish rules for the addition of directed numbers.

However, first we must make some agreements that will enable us to write addition problems without using "added to" to indicate the problem. Our difficulty arises from the fact that the + is used to indicate a positive number and this same sign is used to indicate the operation of ________. Also the – is used to indicate negative number and this same sign is used to indicate the operation of ________.

Agreements

1. A number written without a sign is assumed to be positive. For example "+7" can be written as "7."
2. All numbers meant to be negative must be preceded by the "–" sign.
3. No two signs can be written together unless parentheses are used. For example +6 added to –3 must be written as (+6) + (–3).

EXERCISES

Using the previous agreements, write the following problems in symbols.

1. –6 added to –8 ____________________

2. +8 added to +3 ____________________

3. 7 added to positive 6 ____________________

4. Negative 3 added to 7 ____________________

5. 14 added to –6 ____________________

Exercises continued:

Write in words the meaning of the following symbols.

6. 8 + 6 ______________________________

____

7. (−8) + (−3) ______________________________

____

8. (−8) + (−6) ______________________________

____

Using the ideas of combining numbers on the number line (draw lines if necessary) give the answers to the above problems.

1.

2.

3.

4.

5.

6.

7.

8.

Now let's establish some rules for addition of directed numbers.

First consider a problem like (+3) + (+6).

Here we have two directed numbers with the same direction. The result is therefore going to be a number in the same direction as each of these.

(−3) + (−6) is a similiar problem. Since both numbers are in the same direction the direction of their sum will be the same as each of these.

We can now state a very simple rule for this type of problem: "If the signs of two directed numbers are alike, to add the number we add them without regard to the sign and then put their common sign with the answer."

EXERCISES

Add the following pairs of numbers.

1. +6
   +4

2. −8
   −12

3. +126
   + 15

4. −18
   −7

5. +14
   +12

6. −6
   −3

7. +112
   +214

8. −783
   −141

9. −17
   −112

10. +84
    +108

We now have a rule for numbers with like signs so we must next examine problems involving addition of directed numbers with unlike signs.

Let's look at two of the problems you solved by using a number line.

+6 added to −3

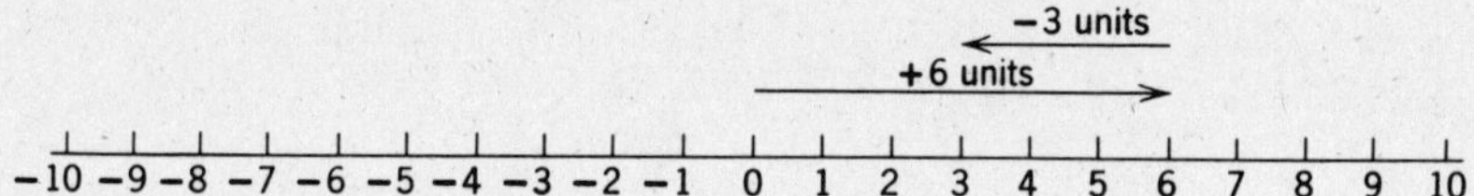

(+6) + (−3) = +3

−6 added to +3

+3 units
−6 units
−10 −9 −8 −7 −6 −5 −4 −3 −2 −1 0 1 2 3 4 5 6 7 8 9 10

(−6) + (+3) = −3

Notice that the answers are the same except for the sign.

Some more examples you worked were:

(+7) + (−12) = −5
(−8) + (+14) = +6
(+8) + (−6) = +2

These problems indicate that the number value of the answer is the difference of the numbers to be added. This is logical since a change in sign means a change in direction.

A little thought will give us the following rule for adding directed numbers with unlike signs: "To add directed numbers with unlike signs, disregard the signs and subtract the smaller number from the larger, then place the sign of the larger number with your answer."

Example: (+7) + (−10)

Subtracting 7 from 10 gives 3 and the sign of the larger number (10) is "−" so (+7) + (−10) = −3

EXERCISES

Add the following pairs of numbers.

1. +6 +2	6. −16 + 3	11. −8 −3
2. −7 −3	7. −18 +18	12. +3 −8
3. +6 −2	8. −9 −11	13. −684 +212
4. −7 +3	9. +6 −14	14. +12 −36
5. +18 −24	10. −16 +15	15. −17 + 3

EXERCISES

Remember the rules regarding punctuation in working the following problems.

1. $[(+6) + (-3)] + (+2) =$ __________
2. $[(-8) + (-12)] + (-4) =$ __________
3. $(-6) + (+14) =$ __________
4. $14 + (-3) =$ __________
5. $17 + 4 =$ __________
6. $17 + (-4) =$ __________
7. $[(-12) + (3)] + 6 =$ __________
8. $[(-9) + (12)] + 3 =$ __________
9. $[(15) + (6)] + (-18) =$ __________
10. $[16 + (-34)] + (+14) =$ __________
11. $-8 + (+12) =$ __________
12. $(14 + 44) + (-50) =$ __________
13. $(163) + [(-6) + (-115)] =$ __________
14. $[(-184) + (-26)] + 200 =$ __________

SUBTRACTION OF DIRECTED NUMBERS

What does subtraction mean to you? ______________________________

____

The concept of subtraction that we will use is of great importance. We will think of subtraction as *finding the difference* between two numbers; or, in terms of a number line, subtraction will mean the directed distance from one number to another.

If the following fact is not already fixed firmly in your mind then think of it now: *"to the right" is a positive direction and "to the left" is a negative direction.*

A small problem presents itself at this point. The problem is in reference to the minus sign, which is used for two purposes: to indicate subtraction and to indicate a negative number.

One more question arises because subtraction is not commutative. Remember that this word simply means that (a – b) is not the same as (b – a).

In dealing with directed numbers it is necessary that we agree that *A – B means the directed distance from B to A.*

Examples: (+6) – (+4)

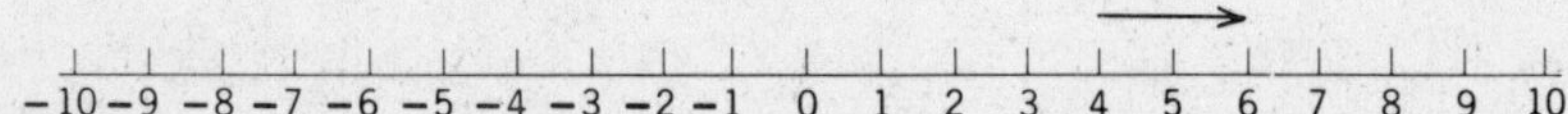

We see that from +4 to +6 is two units to the right. So (+6) – (+4) = +2

(–4) – (–6)

–10 –9 –8 –7 –6 –5 –4 –3 –2 –1 0 1 2 3 4 5 6 7 8 9 10

We see that from (–6) to (–4) is two units to the right. So (–4) – (–6) = +2

EXERCISES

Use the number lines to work the following problems.

1. (−2) −(+4) =

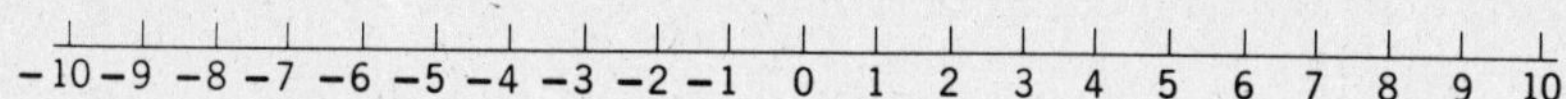

2. (−7) − (−9) =

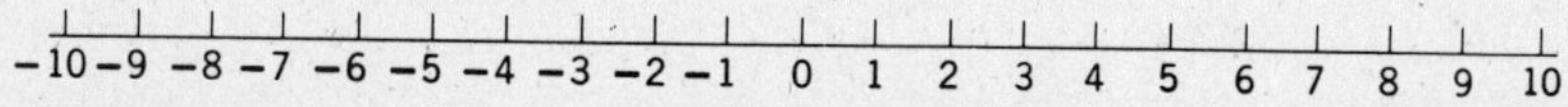

−10 −9 −8 −7 −6 −5 −4 −3 −2 −1 0 1 2 3 4 5 6 7 8 9 10

3. (−9) − (−7) =

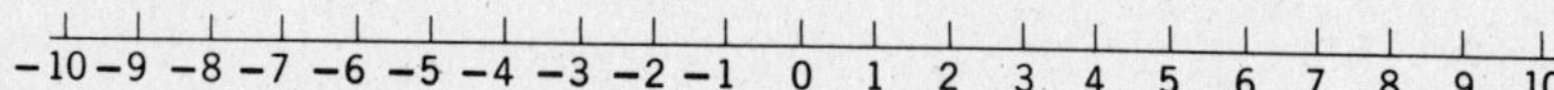

4. (+6) − (−3) =

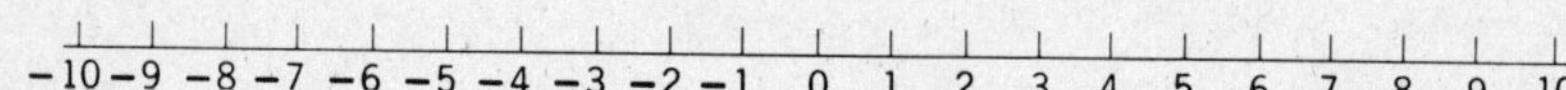

5. (+6) − (−6) =

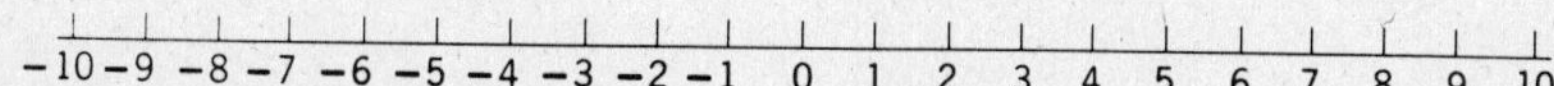

6. (−8) − (−7) =

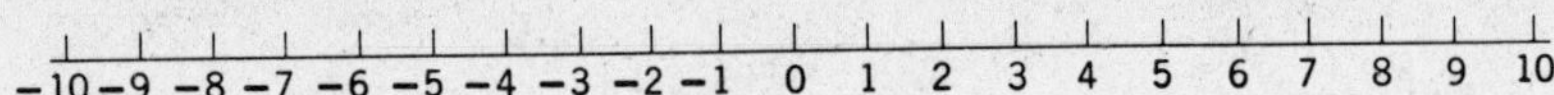

7. (+6) − (+1) =

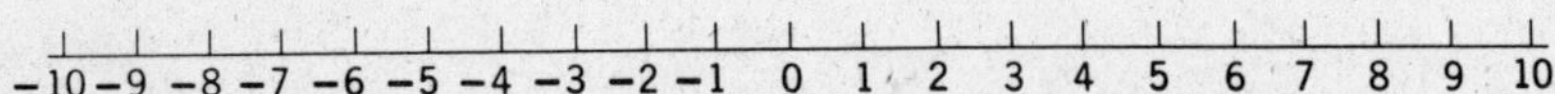

8. (+3) − (+3) =

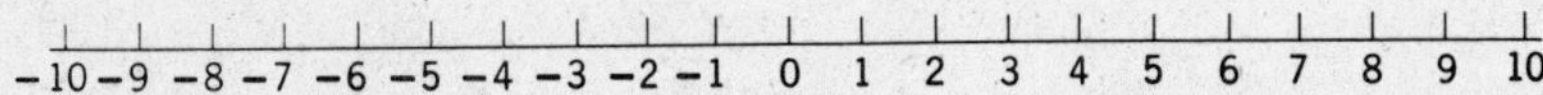

9. (−4) − (+4) =

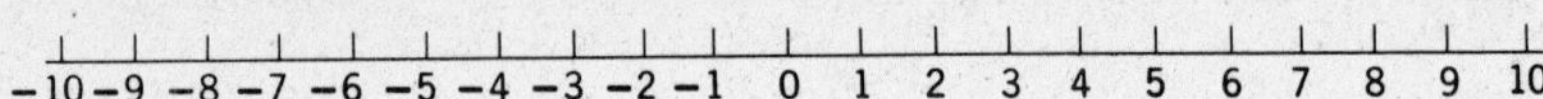

10. (−10) − (+10) =

−10 −9 −8 −7 −6 −5 −4 −3 −2 −1 0 1 2 3 4 5 6 7 8 9 10

EXERCISES

Some of the following problems are addition problems; use the rules we have established to work them. Some are subtraction problems; work them by thinking of the distance from the second number to the first. (Draw a number line if necessary.)

1. (a) $(+6) - (+4) =$ ________ (b) $(+6) + (-4) =$ ________

2. (a) $(-7) - (-3) =$ ________ (b) $(-7) + (+3) =$ ________

3. (a) $(+12) - (-4) =$ ________ (b) $(+12) + (+4) =$ ________

4. (a) $(-3) - (-7) =$ ________ (b) $(-3) + (+7) =$ ________

5. (a) $(-12) - (-4) =$ ________ (b) $(-12) + (+4) =$ ________

6. (a) $(-3) - (-10) =$ ________ (b) $(-3) + (+10) =$ ________

7. (a) $(+6) - (+2) =$ ________ (b) $(+6) + (-2) =$ ________

8. (a) $(14) - (-3) =$ ________ (b) $(14) + (+3) =$ ________

9. (a) $(-10) - (+10) =$ ________ (b) $(-10) + (-10) =$ ________

10. (a) $(+7) - (+7) =$ ________ (b) $(+7) + (-7) =$ ________

Answers to the exercises on the previous page should be the same for parts (a) and (b) of each problem. If we take a close look at these problems we will see that there is a definite relationship between addition and subtraction of directed numbers.

Our rule for subtraction is a simple one: "To subtract one directed number from another, add the opposite of that directed number to the other."

This rule can also be stated another way: "To subtract one directed number from another, change the sign of the number being subtracted and then use the appropriate rule for addition."

Example: $(+6) - (+7) = (+6) + (-7) = -1$

Example: Subtract: $\begin{array}{r} +8 \\ \underline{+6} \end{array}$ Change to $\begin{array}{r} +8 \\ \underline{-6} \end{array}$

and follow the correct rule for addition. The answer is ________.

EXERCISES

1. $(12) - (+8) = (12) + (\_\_) =$ ________

2. $(-8) - (-10) = (-8) + (\_\_) =$ ________

Subtract:

3. $\begin{array}{r} +9 \\ \underline{-6} \end{array}$

4. $\begin{array}{r} +9 \\ \underline{+6} \end{array}$

5. $\begin{array}{r} -6 \\ \underline{-14} \end{array}$

6. $\begin{array}{r} -6 \\ \underline{+14} \end{array}$

7. $\begin{array}{r} -8 \\ \underline{-3} \end{array}$

8. $\begin{array}{r} +8 \\ \underline{-3} \end{array}$

9. $\begin{array}{r} -7 \\ \underline{-16} \end{array}$

10. $\begin{array}{r} +12 \\ \underline{+\ 4} \end{array}$

EXERCISES

1. Subtract

(a) $\begin{array}{r} -7 \\ \underline{-6} \end{array}$ (b) $\begin{array}{r} +7 \\ \underline{-6} \end{array}$ (c) $\begin{array}{r} -7 \\ \underline{+6} \end{array}$ (d) $\begin{array}{r} +7 \\ \underline{+6} \end{array}$

2. Subtract

(a) $\begin{array}{r} -16 \\ \underline{-\ 9} \end{array}$ (b) $\begin{array}{r} +16 \\ \underline{-\ 9} \end{array}$ (c) $\begin{array}{r} -16 \\ \underline{+\ 9} \end{array}$ (d) $\begin{array}{r} +16 \\ \underline{+\ 9} \end{array}$

Do the indicated operations in each of the following.

3. $[(+6) - (-3)] - (+8) =$ ________

4. $[(-16) + (-3)] - (-6) =$ ________

5. $[(+14) - (-6)] - [(+3) + (+7)] =$ ________

6. (18 +7) − (9 + 12) = ________

7. [(+12) − (−6)] + [(−8) − (−9)] = ________

8. [(+16) − (+16)] − [(−8) − (+8)] = ________

9. [8 − (3 + 2)] − 3 + 6 = ________

10. [(−6) − (8 + 3)] − (6 + 4) = ________

Now that we have established rules for addition and subtraction of directed numbers, we will solve some open sentences involving these operations.

Before working with open sentences we must agree about the meaning of an expression like (10 − 7). Does (10 − 7) mean [(+10) − (+7)] or does it mean [(+10) + (−7)]? In other words, is the minus sign an operation sign or is it a negative sign?

If we answer that it is a negative sign, then:

$$(10 - 7) = (+10) + (-7) = +3$$

If we answer that it is an operation sign (i.e., subtraction) then:

$$(10 - 7) = (+10) - (+7) = +3$$

Why does it make no difference which way the sign is interpreted? ________________

____

Example: x − 7 = −3

(a) x − (+7) = −3 — "Seven subtracted from some number is (−3)"; or "the directed distance from 7 to some number is 3 units to the left."

So x = +4

(b) x + (−7) = −3 — The sum of some number and (−7) is −3.

So x = ______

Example: x − 7 = 8

(a) x − (+7) = 8

So x = 15

(b) x + (−7) = 8

So x = _____

Find a replacement value for x in each of the following open sentences so that a true statement will result.

1. $14 - x = 4$

   $x =$ ________

2. $8 + x = -7$

   $x =$ ________

3. $x + 6 = -2$

   $x =$ ________

4. $6 + x = -2$

   $x =$ ________

5. $9 - x = 11$

   $x =$ ________

6. $x - 9 = 11$

   $x =$ ________

7. $14 + (-6) = x$

   $x =$ ________

8. $12 = x - 4$

   $x =$ ________

9. $x + 16 = -3$

   $x =$ ________

10. $3 = x - 7$

    $x =$ ________

11. $x - 4 = 4$

    $x =$ ________

12. $4 + x = 0$

    $x =$ ________

13. $x = 6 - (-8)$

    $x =$ ________

14. $15 - 6 = x$

    $x =$ ________

15. $x + 6 = 15$

    $x =$ ________

16. $x - 15 = -6$

    $x =$ ________

MULTIPLICATION OF DIRECTED NUMBERS

Many rules in mathematics can be discovered by observing patterns or the orderly arrangements that result from some operation.

For example you may have seen a problem like this on some placement test.

"Fill the three blank spaces in this sequence:

2, 4, 7, 11, 16, ___, ___, ___."

In working such a problem you must find the pattern that is extablished by the given numbers. In the foregoing example you would notice that the difference of successive numbers increases by 1 as you move from left to right. So the blanks would contain 22, 29, and 37.

EXERCISES

Discover the pattern used in the following sequences and fill in the blanks.

1. 1, 1, 2, 3, 5, 8, 13, ___, ___, ___

2. 2, 5, 3, 6, 4, 7, 5, ___, ___, ___

3. 10, 8, 6, 4, 2, 0, ___, ___, ___

4. 1, 2, 4, 8, 16, ___, ___, ___

5. −10, −8, −6, −4, −2, 0, ___, ___, ___

6. 1, 2, 4, 5, 7, 8, ___, ___, ___

Let's take a look at possible patterns that develop when we multiply directed numbers and see if we can establish the rules involved.

Consider this set of numbers:

{+5, +4, +3, +2, +1, 0, −1, −2, −3, −3, −5}

Now let's multiply each number in this set by +2. We get {+10, +8, +6, +4, +2, 0 . . .} and then come to the product (+2) × (−1).

Observing the pattern, we see that as we move from left to right in the set, each product decreases by 2 (is 2 less). If this pattern is to continue, the entire set will be:

{+10, +8, +6, +4, +2, 0, −2, __, __, __, __}.

This means that (+2) × (−1) = _____, (+2) × (−2) = _____, (+2) × (−3) = _____, (+2) × (−4) = _____, and (+2) × (−5) = _____.

Now let's take the same set of numbers {+5, +4, +3, +2, +1, 0, −1, −2, −3, −4, −5} and multiply by (−2).

From the previous pattern we can see that (+5) × (−2) should be (−10), and so forth. This gives us the set: {−10, −8, −6, −4, −2, 0, . . .} and we observe that as we move from left to right in the set, each product increases by 2 (is 2 more).

We have now arrived at the product (−2) × (−1). If the pattern is to continue the entire set will be:

{−10, −8, −6, −4, −2, 0, +2, __, __, __, __}.

This means that:

(−2) × (−1) = +2, (−2) × (−2) = ____, (−2) × (−3) = ____,
(−2) × (−4) = ____, and (−2) × (−5) = ____.

If you now take a second look at the two sets of products you should be able to answer the following questions.

What is the sign of the answer when:

(a) two positive numbers are multiplied? ________

(b) two negative numbers are multiplied? ________

(c) a positive and a negative number are multiplied? ________

Can you now state the rule for multiplication of directed numbers? It is: "When multiplying directed numbers, like signs give an answer that is ________, and unlike signs give an answer that is ________."

EXERCISES

Find the products.

1. $(-6)\ (+7) =$ ________

2. $(+2)\ [(-4)\ (-3)] =$ ________

3. $(+6)\ [(-2) + (-3)] =$ ________

4. $(+6)\ (-2) + (+6)\ (-3) =$ ________

5. $(-2)\ (+4)\ (-3)\ (+2) =$ ________

6. $(-2)\ (-4)\ (-3)\ (-2) =$ ________

7. (−2) (−4) (+3) (−2) = ________

8. (−10) [(−8) − (−6) + (+2)] = ________

9. (−3) [(−6) (−4)] = ________

10. (−7) [(+30) + (+4)] = ________

11. (−7) (+34) = ________

12. (9) [7 − 3] = ________

13. (9) (7) − (3) = ________

14. (−2) (−2) (−2) = ________

DIVISION OF DIRECTED NUMBERS

Do you remember the relation between multiplication and division?

We know that $6 \div 3 = 2$ because $2 \times$ ____ = ____. Or, in general terms; $A \div B = C$ if $(C) \times (B) = A$.

Now let's look at some problems with directed numbers.

(a) $(-6) \div (3) = x$. Here we are asking "what number multiplied by 3 will give (−6)?" We know the answer to be a negative number because of our rules of multiplication.

So $(-6) \div (3) =$ ________.

(b) $(-6) \div (-3) = x$. Here a number must be found which can be multiplied by (−3) to give (−6). The answer must be a positive number.

So $(-6) \div (-3) =$ ________.

What does this relationship between multiplication and division tell us about the rules for division? ________________

Complete this rule. "When dividing directed numbers, *like* signs give an answer that is ________________ and *unlike* signs give an answer that is ________________.

EXERCISES

Solve the following problems.

1. $\frac{-24}{+6} =$ ________

2. $\frac{-40}{-8} =$ ________

3. $\frac{+40}{+8} =$ ________

4. $\frac{-18}{-9} =$ ________

5. $2x = -20$

   $x =$ ________

6. $(-3)x = -15$

   $x =$ ________

7. $\frac{-15}{x} = 5$

   $x =$ ________

8. $\frac{-24}{x} = 7$

   $x =$ ________

9. $\frac{+32}{x} = -4$

   $x =$ ________

10. $\frac{-27}{-3} = x$

    $x =$ ________

11. $15x = -60$

$x =$ ________

12. $3\frac{1}{4}x = -13$

$x =$ ________

13. $5x = -40$

$x =$ ________

14. $2\frac{1}{2}x = -7\frac{1}{8}$

$x =$ ________

15. $8x = -72$

$x =$ ________

16. $2x + 4 = 12$

$x =$ ________

17. $-14x = 49$

18. $4x - 10 = 10$

19. $3\frac{1}{2}x - 8 = 6$

20. $-3x + 8 = 17$

21. $-.6x = 3.6$

22. $\frac{3}{4}x + 2 = 11$

23. $.75x - 9 = 3$

24. $-15x = -10$

Chapter 14

IRRATIONAL NUMBERS

We have dealt with several sets of numbers thus far in this text.

We first studied the *counting numbers:* 0, 1, 2, 3, . . .

We then studied the set of *positive fractions:* $\frac{1}{2}, \frac{2}{3}, \frac{5}{11}, \ldots$

These two sets were combined and called the *numbers of arithmetic.*

The numbers of arithmetic are defined as those numbers that can be written as $\frac{a}{b}$, where a and b are counting numbers and b is *not* zero.

Which of these are *not* numbers of arithmetic?

$$0, \frac{3}{4}, \frac{0}{6}, -12, 1.7, \frac{6}{11}, \frac{5}{3}, -2, \frac{5}{8}, \frac{4}{0}$$

____

We next studied the *negatives* of the numbers of arithmetic.

A set we have used but not defined formally is the set of *integers.* The set of integers is made up of the set of counting numbers and their negatives.

Which of the following are *not* integers?

$$0,\ -6,\ \frac{2}{3},\ 5,\ \frac{0}{6},\ \frac{5}{8},\ \frac{10}{2},\ -\frac{6}{3},\ -4$$

____

Notice that the set of counting numbers is *included* in the set of integers.

The numbers of arithmetic and their negatives make up a set called the set of *rational numbers.*

Is a counting number a rational number? ________

Is an integer a rational number? ________

Is a number of arithmetic a rational number? ________

Is the negative of a number of arithmetic a rational number? ________

Since the numbers of arithmetic are defined in terms of the counting numbers and since the rationals include the numbers of arithmetic and their negatives, we can define the rationals in terms of the counting numbers and their negatives — the integers.

A rational number is any number that can be written as $\frac{a}{b}$, where a and b are integers and b is not zero.

Which of the following are *not* rational numbers?

$$-6,\ \frac{2}{3},\ 0,\ \frac{5}{8},\ \frac{2}{7},\ \frac{7}{2},\ 4.6,\ \frac{5}{0}$$

____

EXERCISES

1. Show that $\frac{\frac{2}{3}}{\frac{7}{9}}$ is a rational number. (i.e., find a form $\frac{a}{b}$ for $\frac{\frac{2}{3}}{\frac{7}{9}}$)

2. Show that 27.6 is a rational number.

3. Show that $-5\frac{1}{6}$ is a rational number.

4. Show that 2 is a rational number.

Can each of these rational numbers be expressed in more than one way? ____________

RATIONAL NUMBERS AS DECIMALS

Change these rational numbers to decimals.

1. $\frac{2}{5}$ = ____________

2. $\frac{3}{100}$ = ____________

3. $\frac{6}{20}$ = ____________

4. $2\frac{7}{8}$ = ____________

5. $\frac{2}{3}$ = ____________

This last problem cannot be expressed exactly as a decimal without some special notation.

We can say that $\frac{2}{3} = .666\ldots$ (with the three dots indicating that the sixes would continue forever) or we can write $\frac{2}{3} = .66\overline{6}$ (where the bar over the 6 indicates that the sixes will continue).

.234234234 . . . should indicate that the 234 will repeat. However there is room for wondering just what will repeat. A better notation is $.234\overline{234}$, where the bar is used to indicate the numbers that repeat.

The set of numbers that repeat is called a *period.*

In this notation a bar over a period is used to indicate a *repeating decimal.*

Some examples of repeating decimals are:

(a) $.621\overline{621}$ (b) $.385\overline{85}$

(c) $.214\overline{343}$ (d) $.286\overline{6}$

Notice that a period does not necessarily include all of the numbers.

Now let's see where repeating decimals might occur.

Example: Change $\frac{3}{7}$ to a decimal.

```
    .42857142
7 | 3.00000000
    2 8
     20
     14
      60
      56
       40
       35
        50
        49
         10
          7
          30
          28
           20
           14
            6...
```

Do you see why we can write $\frac{3}{7} = .428571\overline{428571}$?

Change the following rational numbers to decimals. If they repeat, write them, using the bar over the period.

1. $\frac{24}{99}$ 2. $\frac{137}{990}$

Will all rational numbers give a repeating period when changed to decimal notation?

To answer this question let's look back to our first example: change $\frac{3}{7}$ to a decimal. We divide 7 into 3 by adding a decimal point and as many zeros as we wish. When we divide by 7 what are our possible remainders? Could we have a remainder of 8? of 7? List the possible remainders.

____ ____ ____ ____ ____ ____ ____

You should have listed 0 through 6 as possible remainders.

If we have only seven possible remainders when dividing by 7 then we must soon get a remainder that has occurred before.

If we have a remainder of zero then our division has "come out even." This means that $\frac{3}{7}$ as a decimal must either come out even or repeat a remainder within _____ places.

Suppose we wish to express $\frac{6}{13}$ in decimal form. This number would either come out even or repeat within the first _____ places.

This leads us to realize that every number that can be expressed as $\frac{a}{b}$ will either come out even or repeat within b places when expressed as a decimal.

Example: Change $\frac{7}{8}$ to a decimal.

We know that this number will come out even or repeat within _____ places.

$$\begin{array}{r} .8750 \\ 8\overline{)7.0000} \\ \underline{6\,4} \\ 60 \\ \underline{56} \\ 40 \\ \underline{40} \\ 0 \end{array}$$

So $\frac{7}{8} = .875$

We could write $\frac{7}{8} = .875\overline{0}$, where the repeating period is 0. We can now state this fact: "Any rational number can be expressed as a repeating decimal."

EXERCISES

Express each of the following numbers as repeating decimals.

1. $\frac{1}{7}$ = ______________

2. $\frac{3}{4}$ = ______________

3. 5 = ______________

4. $\frac{3}{11}$ = ______________

5. $-\frac{4}{5} =$ _______________

6. $2\frac{1}{3} =$ _______________

7. $\frac{51}{99} =$ _______________

Because we see that all rational numbers can be expressed as repeating decimals we might wonder if all repeating decimals are rational numbers.

Example: Change $.375\overline{0}$ to a rational number in lowest terms.

$$.375\overline{0} = \frac{375}{1000} = \frac{3}{8}$$

This example with the zero repeating is a type that is easily changed to a rational number.

Change the following to rational numbers in lowest form.

1. $6.25\overline{0}$ = ______________
2. $.875\overline{0}$ = ______________
3. $.6\overline{0}$ = ______________
4. 2.6 = ______________
5. .231 = ______________
6. .125 = ______________

Can we multiply a repeating decimal by another number?

Example 1: Multiply $.33\overline{3}$ by 2.

We should probably write $.33\overline{3}$ as .333 . . . to see that multiplying by 2 would be the same as .333 . . . + .333 . . . or

$$\begin{array}{r} .333\ldots \\ +\ .333\ldots \\ \hline .666\ldots \end{array}$$

Example 2: Multiply $.33\overline{3}$ by 10.

This example is easier because multiplying by 10 simply moves the decimal point ______ places to the ________.

$$.33\overline{3} \times 10 = 3.33\overline{3}$$

EXERCISES

Perform the indicated operation.

1. $.66\overline{6} \times 2 =$ __________

2. $.66\overline{6} \times 100 =$ __________

3. $5.63\overline{63} \times 100 =$ __________

4. $.743\overline{43} + 3.51\overline{51} =$ __________

5. $1.26\overline{26} - .26\overline{26} =$ __________

Now let's try changing a repeating decimal, with a period other than zero, to a rational number.

Example 1. Change $.66\overline{6}$ to a rational number.

Step 1. Let $x = .66\overline{6}$.
Here we are saying that some unknown rational number is $.66\overline{6}$.

Step 2. If x and $.66\overline{6}$ are the same numbers then 10 times x and 10 times $.66\overline{6}$ are the same numbers. So $10x = 6.66\overline{6}$.

Step 3. Subtract $x = .66\overline{6}$ from $10x = 6.66\overline{6}$.

$10x - x = 9x$ and $6.66\overline{6} - .66\overline{6} = 6$.

So $9x = 6$,

and $x = \frac{6}{9} = \frac{2}{3}$;

therefore $.66\overline{6} = \frac{2}{3}$.

Example 2: Change $.14\overline{14}$ to a rational number.

Let $x = .14\overline{14}$.

Multiply by 100. $100x = 14.14\overline{14}$.

Subtract the first equation from the second.

$99x = 14$.

$x = \frac{14}{99}$

So $.14\overline{14} = \frac{14}{99}$

Example 3: Change $.23\overline{14}$ to a rational number.

Let $x = .23\overline{14}$.

Multiply by 100. $100x = 23.14\overline{14}$.

Subtract:

$$\begin{array}{r} 23.14\overline{14} \\ -\ \ .23\overline{14} \\ \hline 22.91 \end{array}$$

So $99x = 22.91$

$$x = \frac{22.91}{99} = \frac{2291}{9900}$$

Notice that in some examples we multiplied by 10 and in some by 100. What determines the number we wish to use as a multiplier? ______________________

Your answer should be *the length of a period,* since we wish to subtract in such a way that we will get zeros beyond a certain point.

If we multiply by a number that will move the decimal point a complete period, then the repeating period will be lined up when we subtract.

Example 4: Change $.231\overline{231}$ to a rational number.

Since the period is three digits in length we will multiply by 1000.

Let $x = .231\overline{231}$

$1000x$ = ________

$999x$ = ________

x = ________

EXERCISES

Change the following to rational numbers.

1. $.75\overline{75}$ = ______________

2. $.321\overline{21}$ = ______________

3. $.27\overline{27}$ = ______________

4. $.107\overline{07}$ = ______________

5. $.571428\overline{571428}$ = ______________

6. $.246\overline{46}$ = ____________

7. $.232\overline{232}$ = ____________

8. $.12\overline{12}$ = ____________

9. $.503\overline{3}$ = ____________

10. $.12\overline{2}$ = ____________

We have now established two important facts: Every rational number can be expressed as a repeating decimal, and Every repeating decimal can be expressed as a rational number.

Do numbers exist that are not rational? If so, they cannot be repeating decimals.

To show that numbers that are not rational actually exist we must exhibit a decimal that *does not repeat.*

Try to find a repeating period in these decimals.

(a) .25225222522225 . . .

(b) .01001000100001 . . .

(c) .335533355533335555 . . .

(d) .012345678900123456789000 1 . . .

Obviously numbers exist that are not rational. They are called the *irrational numbers.*

EXERCISES

Write four irrational numbers.

1. ____________________

2. ____________________

3. ____________________

4. ____________________

Perhaps the most widely used irrational numbers are those that are *square roots* of other numbers.

Remember that the square of a number is found by using that number as a factor two times.

The square of 12 is 144 since (12) (12) = 144. The square of $3\frac{1}{2}$ is $12\frac{1}{4}$ because: ____________________________. We know that $12 \cdot 12$ can be written as 12^2; $4 \cdot 4$ can be written as 4^2; and (−7) (−7) is written as (_____)^2. The exponent 2 indicates that the number is used as a factor twice.

Evaluate the following numbers.

1. 3^2 = ________________
2. $(1.7)^2$ = ________________
3. $(-19)^2$ = ________________
4. $(\frac{3}{4})^2$ = ________________
5. 0^2 = ________________
6. $(3\frac{2}{3})^2$ = ________________
7. 14^2 = ________________
8. $(.0061)^2$ = ________________
9. $(.14)^2$ = ________________
10. 16.3^2 = ________________
11. 13^2 = ________________
12. $(2 \cdot 3)^2$ = ________________
13. $(1\frac{1}{2})^2$ = ________________
14. 19^2 = ________________
15. $(\frac{3}{4} + \frac{1}{4})^2$ = ________________
16. $(.2)^2$ = ________________
17. 11^2 = ________________
18. $(-11)^2$ = ________________
19. $(\frac{1}{2})^2$ = ________________
20. $(12\frac{1}{2})^2$ = ________________

Squares of counting numbers are called *perfect squares.* Some of these are 0, 1, 4, 9, 16, . . . since they are squares of 0, 1, 2, 3, 4, ...

In the column at the right list the perfect squares through 400.

____

____

____

____

____

____

____

____

____

____

____

____

____

____

____

____

____

____

____

____

____

Factor the following numbers into prime factors.

1. 36 = ________
2. 225 = ________
3. 144 = ________
4. 24 = ________
5. 32 = ________
6. 48 = ________

Which of the above numbers are perfect squares? Problems ____, ____, and ____.

Now look at the prime factors of problems 1, 2, and 3 and compare them with problems 4, 5, and 6. You should notice that all exponents in problems 1, 2, and 3 are ________ (odd, even) numbers, whereas in problems 4, 5, and 6 at least one exponent is ________ (odd, even).

Remember this rule: "All perfect square counting numbers will factor into primes such that each exponent will be even."

Try a few more numbers, both perfect squares and those that are not perfect squares, to convince yourself that this rule is true.

1. 6 = ________
2. 120 = ________
3. 441 = ________
4. 289 = ________
5. 900 = ________
6. 153 = ________

Which of these numbers are perfect squares?

1. $2^4 \cdot 3^2 \cdot 5^4$

2. $3^2 \cdot 7^2 \cdot 11^4$

3. $2^3 \cdot 3^4 \cdot 5^2$

4. 43^3

5. $17^2 \cdot 2^3$

Example: By what must $2^3 \cdot 3^2 \cdot 5^5$ be multiplied to make it a perfect square?

Answer: If 2^3 is multiplied by 2 it will be 2^4 and have an even exponent, and if 5^5 is multiplied by 5 it will be ______ and have an even exponent. So $2^3 \cdot 3^2 \cdot 5^5$ must be multiplied by $2 \cdot 5$ to give $2^4 \cdot 3^2 \cdot 5^6$, which is a perfect square.

By what must each of the following be multiplied to give a perfect square?

1. $2^3 \cdot 3 \cdot 5^2$ ________________

2. $2^2 \cdot 3^3 \cdot 5^3$ ________________

3. $5^2 \cdot 7 \cdot 11$ ________________

4. $3^3 \cdot 2^3 \cdot 5^3$ ________________

Example 1. What is the largest perfect square that is a factor of 150?

$150 = 2 \cdot 3 \cdot 5^2$ in prime factored form.

So $5^2 = 25$ is the largest perfect square factor.

Example 2. What is the largest perfect square factor of 96?

$96 = 2^5 \cdot 3$ in prime factored form.

So $2^4 = 16$ is the largest perfect square factor.

EXERCISES

Find the largest perfect square factor of the following. (First find the prime factored form of each number.)

1. 84 ______________________________ ____________

2. 12 ______________________________ ____________

3. 98 ______________________________ ____________

4. 75 ______________________________ ____________

5. 216 ______________________________ ____________

6. 112 ______________________________ ____________

7. 1176 ______________________________ ____________

Frequently we must reverse the operation of squaring a number. That is, we are given a number (call it A) and asked to find all numbers that when squared will equal A.

These are called *square roots* of A. Every positive number has *two* square roots.

The two square roots of 361 can be found in problems 3 and 14 in the exercise on page 214. They are ______ and ______. We find from problems 17 and 18 that the two square roots of 121 are ______ and ______.

List pairs of square roots for these numbers:

4:	______	______	.0081:	______	______
25:	______	______	144:	______	______
100:	______	______	1:	______	______
400:	______	______	$(\frac{2}{3})^2$:	______	______
169:	______	______	225:	______	______
$\frac{1}{4}$:	______	______	$12\frac{1}{4}$:	______	______
$\frac{1}{16}$:	______	______	4^2:	______	______
.09:	______	______	$20\frac{1}{4}$:	______	______
.16:	______	______	$(2 \cdot 3)^2$:	______	______
289:	______	______	$(11.37)^2$:	______	______
36:	______	______	196:	______	______
81:	______	______	a^2:	______	______
$(-7)^2$:	______	______	256:	______	______

All of the numbers given above are positive numbers. We cannot consider square roots of negative numbers since each time we square a number (whether positive or negative) the result is positive.

A common symbol for the positive square root of a number is the *radical,* $\sqrt{}$ written over the number.

The positive square root of 144 is written: $\sqrt{144}$ This is the same number as +12. The positive square root of 49 is written $\sqrt{49}$, This is equal to ______.

Remember that 49 has two square roots. They are +7 and −7; but $\sqrt{49}$ is a symbol for the positive root only, which is +7 in this case.

EXERCISES

Mark each of the following either True or False.

1. ______ $\sqrt{9} = 3$
2. ______ $\sqrt{169} = +13$
3. ______ $\sqrt{25} = -5$
4. ______ $\sqrt{4} = -2$
5. ______ $\sqrt{400} = 20$
6. ______ $\sqrt{600} = 30$
7. ______ $\sqrt{.09} = -.3$
8. ______ $\sqrt{1/2} = 1/4$
9. ______ $\sqrt{25/36} = -5/6$
10. ______ $\sqrt{1000} = 100$
11. ______ $\sqrt{-4} = -2$
12. ______ $\sqrt{16} = 4$
13. ______ $\sqrt{16} + \sqrt{25} = 9$
14. ______ $\sqrt{36} - \sqrt{4} = \sqrt{32}$
15. ______ $\sqrt{55}$ is between 7 and 8
16. ______ $\sqrt{10}$ is between 3 and 4
17. ______ $\sqrt{32/2} = 4$

Zero also has two square roots. They are ______ and ______, but both of these equal ______. Therefore,

$$\sqrt{10-10} = \_\_\_\_\_\_, \quad \sqrt{0} = \_\_\_\_\_\_, \quad \text{and} \quad \sqrt{a-a} = \_\_\_\_\_\_.$$

Square roots or approximate square roots can be found for any number written in decimal form. If the square root is rational it can be found exactly; otherwise it must be approximated. This approximation can be determined to any degree of accuracy that we wish. Usually it is sufficient to determine a square root to two or three decimal places, that is, correct to hundredth or thousandth.

The following examples illustrate a "division" method for finding these square roots.

Example 1: Find $\sqrt{1849}$.

First group the digits in pairs moving to the left from the decimal point.

$$\sqrt{\underline{18}\,\underline{49}}$$

Consider next the leftmost pair, which is 18. Select the largest perfect square that does not exceed 18. In this case it is 16. The first digit of our answer will be the positive square root of 16, which is 4. The digit 4 is placed above the first *pair* and 16 is placed below the first pair.

Subtract, as below and "bring down" the next pair of numbers.

```
   4   .
 √18 49.
  16
  ─────
   2 49
```

Next, double whatever number appears above 1849. In this case the result is 8. (4 doubled) Place the number 8 to the left of 249 and to the right of 8 place a small dash as a place holder in the units position. Our problem now looks like this:

```
      4   .
    √18 49.
     16
     ──────
 8_ │ 2 49
```

. . . and 249 is to be divided by "eighty – something." That is we try the divisors 81, 82, 83, 84, ... In this problem the 8$\underline{3}$ is the correct divisor. The number $\underline{3}$ is placed

1. in the position to the right of 8, and
2. above the next group of numbers. (In this case, above the 49.)

Multiply 3 times 83 and place the product under 249, as follows:

```
      4  3.
    √18 49.
     16
     ──────
 83 │2 49
    │2 49
     ──────
         0
```

The zero remainder indicates that 1849 is a perfect square and 43 is the positive square root.

Example 2: Find the square root of 198.63 correct to hundredths.

$$\sqrt{\underline{1}\ \underline{98}\,.\,\underline{63}\ \underline{00}}$$

First mark off the pairs of numbers beginning at the decimal point and moving to the left, then to the right. Notice that two zeros were added in order to have a pair of numbers corresponding to the hundredths position of the answer.

Also notice that the left-most "pair" contains only the single number, 1. To complete the pair we could put a 0 to the left of 1.

Next, select the largest perfect square that does not exceed one. (It is, of course, 1.) Place this below the left pair and the square root of 1, above the pair. Subtract; bring down the next pair.

```
       1
     √01 98 . 63 00
       1
      ----
        98
```

Then, double the 1 (which appears in the answer space). Place this (2) to the left of 98 with the dash indicating that we shall divide by "twenty—something." Twenty-four will do since 4 × 24 is less than 98. Thus:

```
       1  4 .
     √01 98 . 63 00
       1
      ----
 24    |98
       |96
       ----
         263
```

We have subtracted and brought down the next pair.

Next, double 14 and our new divisor will be "two hundred eighty—something."

```
       1  4 .  0
     √01 98 . 63 00
       1
      ----
 24    |98
       |96
       ----
 280     |263
         |000
```

This number is too large to divide 263. A zero is placed above the 63 and also in the space to the right of 28. Multiplication by zero gives the foregoing result.

Subtract again. Bring down the next pair. Double the 140 and place the dash to the right. Nine is the correct entry in this case. It is placed above the 00 and to the right of 280_.

```
            1  4 .  0  9
         √01 98 . 63 00
            1
24          98
            96
280          263
             000
2809         26300
             25281
              1019
```

After multiplication and then subtraction we notice the remainder to be so small with respect to the divisor 2809 that we feel sure that we now have the square root correct to hundredths.

Complete the following and check your answers by squaring.

```
1.      7 .
      √53 . 29
       49
   14_ | 429
```

```
2.      6  7 .
      √45 15 . 84
       36
   12_ | 915
```

3. $\sqrt{3.000000}$ (partial root: 1. 7)

1

27̲ | 200

(Answer correct to 3 decimal places should be 1.732.)

4. $\sqrt{26569.}$ (partial root: 1 .)

1

26̲ | 165

5. $\sqrt{30.0000}$ (partial root: 5)

25

(Round off correct to tenths.)

Determine the positive square roots of these numbers, using the method of the previous problems. Answers correct to tenths will be sufficient.

1. $\sqrt{11}$ = ________ 2. $\sqrt{7.}$ = ________

3. $\sqrt{2916}$ = ________ 4. $\sqrt{.81}$ = ________

5. $\sqrt{45667.69}$

6. $\sqrt{6}$

7. $\sqrt{75}$

8. $\sqrt{8.00}$

9. $\sqrt{12.96}$

10. Complete this table. Use results of the foregoing problems. Notice that most of these square roots are not exact. Better approximations could be obtained by continuing our "division" process and rounding off the result to hundredths, thousandths, etc.

N	$\sqrt{N}$
1	1
2	1.4
3	
4	
5	2.2
6	
7	
8	
9	3
10	3.2
11	
12	3.4

SIMPLIFYING RADICALS

Example 1: $\sqrt{4}$ = ________ $\sqrt{9}$ = ________

$\sqrt{4} \times \sqrt{9}$ = ____ x ____ = _____

$\sqrt{4 \times 9} = \sqrt{\_\_\_\_}$ = _____

Example 2: $\sqrt{16}$ = ________ $\sqrt{25}$ = ________

$\sqrt{16} \times \sqrt{25}$ = ____ x ____ = _____

$\sqrt{16 \times 25} = \sqrt{\quad}$ = ________

These two examples represent a very important fact:

$\sqrt{A \times B} = \sqrt{A} \times \sqrt{B}$ where A and B are any positive numbers.

EXERCISES

Using this rule, evaluate the following.

1. $\sqrt{8} \times \sqrt{2} = \sqrt{\_\_\_\_}$ = ________

2. $\sqrt{5} \times \sqrt{20} = \sqrt{\_\_\_\_}$ = ________

3. $\sqrt{6} \times \sqrt{6} = \sqrt{\_\_\_\_}$ = ________

4. $\sqrt{9} \times \sqrt{4} = \sqrt{\_\_\_\_}$ = ________

5. $\sqrt{12} \times \sqrt{3} = \sqrt{\_\_\_\_}$ = ________

The fact that $\sqrt{A \times B} = \sqrt{A} \times \sqrt{B}$ can be used to simplify some radicals. A radical is in simplified form if the number under the radical does *not* contain a perfect square factor.

Example 1: Simplify $\sqrt{48}$.

First we factor 48 into primes:

$\sqrt{48} = \sqrt{2^4 \cdot 3}$

Next recall that even powers indicate perfect squares; so we wish now to write:

$\sqrt{48} = \sqrt{2^4 \cdot 3} = \sqrt{2^4} \times \sqrt{3}$

but $\sqrt{2^4} = 4$ So $\sqrt{48} = \sqrt{2^4} \times \sqrt{3} = 4 \times \sqrt{3}$

Example 2: Simplify $\sqrt{1176}$.

$$\begin{aligned} \sqrt{1176} &= \sqrt{2^3 \times 3 \times 7^2} \\ &= \sqrt{2^2 \times 2 \times 3 \times 7^2} \\ &= \sqrt{2^2} \times \sqrt{7^2} \times \sqrt{2 \times 3} \\ &= 2 \times 7 \times \sqrt{6} \\ &= 14\sqrt{6} \end{aligned}$$

Simplify each of the following.

1. $\sqrt{24}$

2. $\sqrt{40}$

3. $\sqrt{50}$

4. $\sqrt{288}$

5. $\sqrt{243}$

6. $\sqrt{500}$

7. $\sqrt{360}$

8. $\sqrt{242}$

USING SIMPLIFIED RADICALS

In a previous exercise you made a list of square roots of the numbers 1–12. Let's see now how this list can be used.

Example: Find the square root of 48 correct to two decimal places.

$$\sqrt{48} = \sqrt{4^2 \times 3} = \sqrt{4^2} \times \sqrt{3} = 4 \times 1.73 = 9.62$$

Example: Find the square root of 30 correct to two places.

$$\sqrt{30} = \sqrt{2 \times 3 \times 5} = \sqrt{2} \times \sqrt{3} \times \sqrt{5}$$
$$= 1.41 \times 1.73 \times 2.24$$
$$= 5.46 \text{ rounded to two places.}$$

This multiplication may be considered easier than using the square root algorithm.

EXERCISES

Refer to the list of square roots of the numbers from 1–12 and find the following correct to two decimal places.

1. $\sqrt{50}$
2. $\sqrt{40}$
3. $\sqrt{288}$
4. $\sqrt{1176}$
5. $\sqrt{500}$
6. $\sqrt{360}$
7. $\sqrt{242}$
8. $\sqrt{243}$

THE PYTHAGOREAN THEOREM

An angle of 90° (ninety degrees) is called a *right angle.* Right angles are used more in our every day life than any other angle. For instance: two walls of a room meet at a right angle; the floor of a room meets a wall at a right angle; many intersections of roads are at right angles; the yard-lines of a football field meet the sidelines at right angles; and so forth. You can find many places where right angles occur around you.

Suppose you had the problem of constructing a right angle. What would you do? Perhaps you say that you would simply guess and maybe you could guess pretty close. But if the task is to construct a right angle exactly – say for determining a property line – then guesswork is not enough.

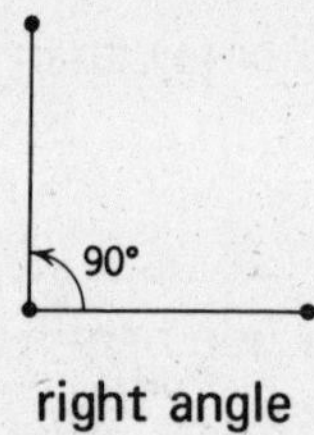

right angle

The ancient Egyptians were faced with this problem. Their lands were flooded every year by the Nile River and all property lines were destroyed. In order for a person to reclaim his land after each flood there had to be some way of correctly constructing a right angle.

Of course, with our modern surveying instruments, this is no problem. For the early Egyptian, however, it was a very serious problem. At some time in this early age the Egyptians made an important discovery. They found that the lengths of 3 units, 4 units and 5 units as the sides of a triangle would always produce a right angle.

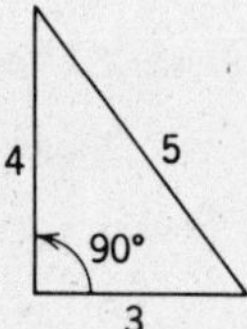

This ratio of sides always gives a right angle.

The early Egyptian surveyors used ropes with 13 equally spaced knots to find a right angle. If you were given a rope with thirteen knots in it, all equally spaced, explain how you would find a right angle.

____

____

Many years after the Egyptians had first used this method of finding right angles, a Greek mathematician named Pythagoras became aware of the fact that the numbers used by the early Egyptians had a special relationship. The Egyptians had used any ratio of 3 to 4 to 5.

Some such ratios are:

(a) 3 4 5 (c) 9 12 ___

(b) 6 8 10 (d) 12 ___ ___

Notice that $3^2 + 4^2 = 5^2$

or $9 + 16 = 25$

Check this relationship with (b), (c), and (d) above.

(b) ____________

(c) ____________

(d) ____________

Pythagoras then proved that in any right triangle (a triangle with a right angle) the square of the length of the longer side is equal to the sum of the squares of the lengths of the other two sides.

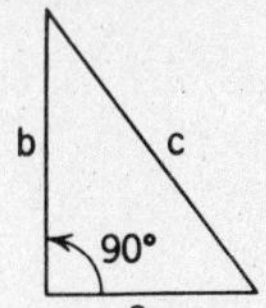

If a, b, and c are the measures of the sides of this right triangle, then $c^2 = a^2 + b^2$.

The longest side of a right triangle (the side opposite the 90° angle) is called the *hypotenuse.* So a correct statement of the Pythagorean theorem is: "The square on the hypotenuse of a right triangle is equal to the sum of the squares on the other two sides."

Example 1: Find the missing side of this right triangle.

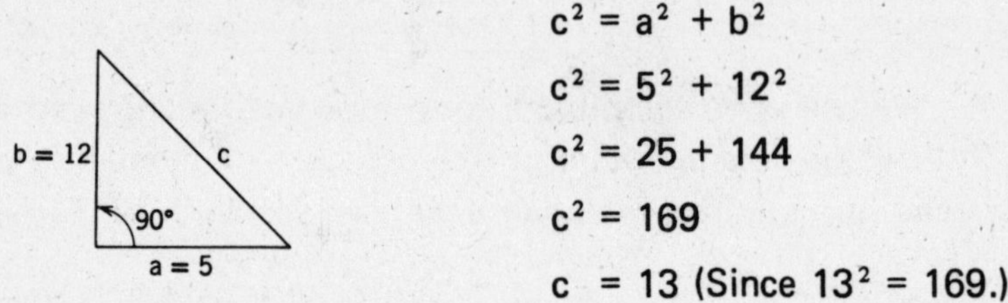

$$c^2 = a^2 + b^2$$
$$c^2 = 5^2 + 12^2$$
$$c^2 = 25 + 144$$
$$c^2 = 169$$
$$c = 13 \text{ (Since } 13^2 = 169.)$$

Note: Distances are always positive numbers, therefore we do not use the other square root of 169.

Example 2: Find the missing side of this triangle.

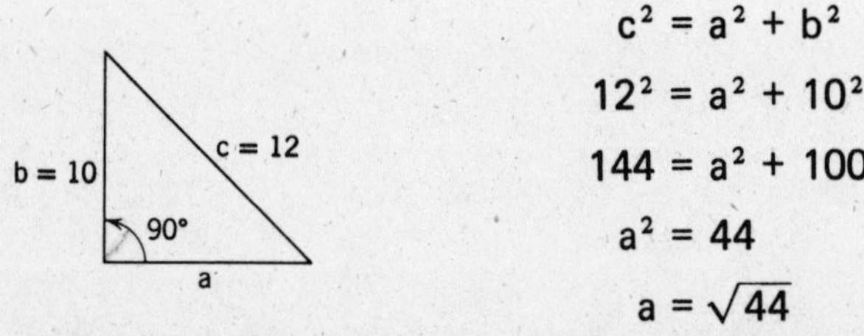

$$c^2 = a^2 + b^2$$
$$12^2 = a^2 + 10^2$$
$$144 = a^2 + 100$$
$$a^2 = 44$$
$$a = \sqrt{44}$$

To find this number correct to one decimal place you can use either of the methods learned earlier. The easier way is to simplify the radical first.

$a = \sqrt{44} = \sqrt{4 \times 11} = \sqrt{4} \times \sqrt{11}$

$= 2x$__________ (refer to the table you completed on page 227)

$a =$ ________.

PERIMETER □ = 4W, 2L + 2W = P▭

EXERCISES

If "c" is the hypotenuse and "a" and "b" are the other two sides, find the missing side.

1. a = 4
   b = 8
   c = ___

2. a = ___
   b = 24
   c = 25

3. a = 10
   b = 12
   c = ___

4. a = 1
   b = 3
   c = ___

EXERCISES

Draw a picture of these problems to help you decide which value is missing.

1. A light pole is 40 feet high and a guy wire is stretched from the top of the pole to a point on the ground 14 feet from the base of the pole. How long is the wire?

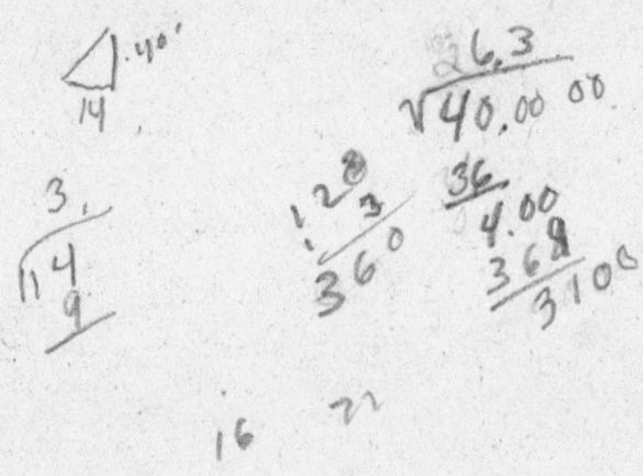

2. A baseball "diamond" is a square (90° angles at each corner) 90 feet on a side. How far is it from home plate to 2nd base? (correct to one decimal place)

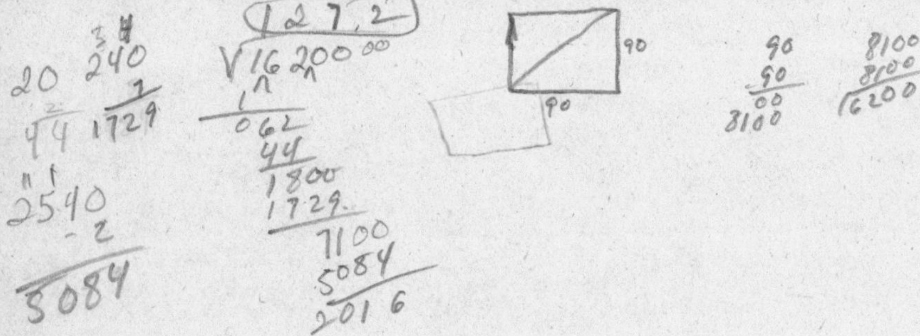

3. The pilot of a plane records his altitude as 1 mile when he is directly over a point 9 miles from his landing point. What is the distance from the plane to the landing point?

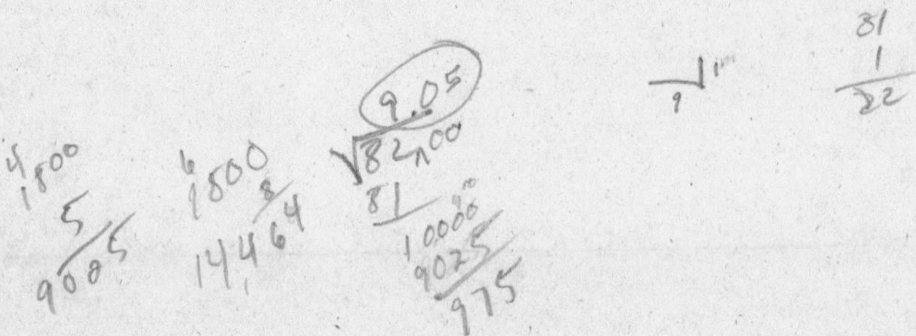

4. A car travels from point A to point B by going 20 miles north and 60 miles west. How much less would the distance have been if the car could have gone directly from point A to point B? (Correct to tenths.)

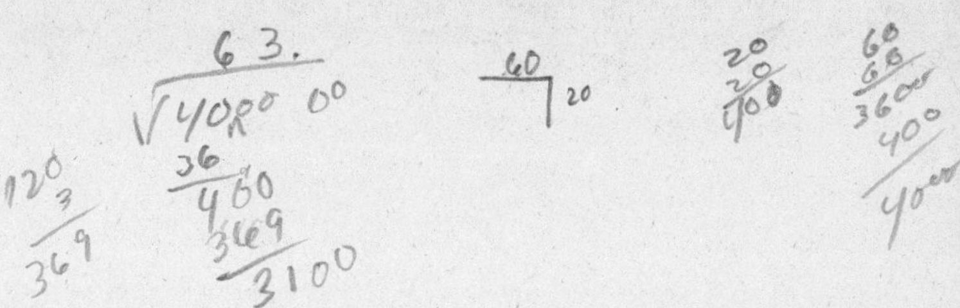

5. Students decorating the gym for a school dance wished to stretch a wire from a point 16 feet hich on the wall to a point on the floor 18 feet from the base of the wall. How much wire do they need? (correct to tenths)

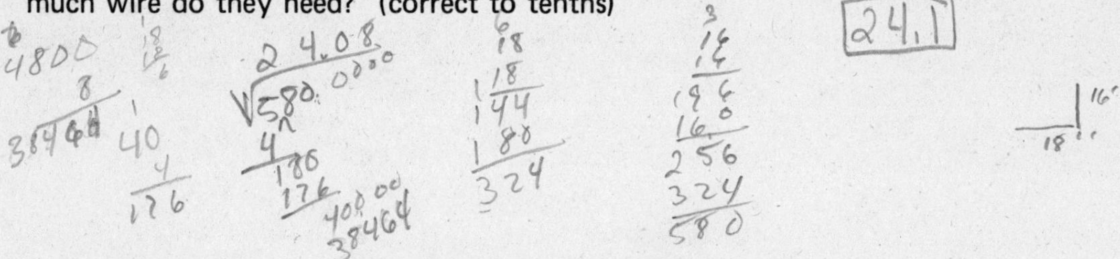

6. What is the distance (in yards correct to tenths) diagonally across a football field? (length: 100 yds; width: 160 ft.)

7. If one leg of a right triangle is $\sqrt{18}$ inches and the other leg is 15 inches, how long is the hypotenuse? Express the answer first in the form of $a\sqrt{b}$ and then find it correct to two decimal places.

8. A farmer has a field in the shape of a square, 100 feet on a side. He wishes to put a fence around the field and a fence along a diagonal. How many feet of fence will he need? (correct to tenths)

9. A light pole is 40 feet high and a guy wire 60 feet long is stretched from the top of the pole to a point on the ground. How far away from the foot of the pole is the guy wire anchored?

Chapter 15

THE LANGUAGE OF ALGEBRA

The use of letters to represent numbers is a principal characteristic of algebra. An example of this is found in a common business formula, $A = P(1 + r)^n$. Here the letter A represents the *amount* (number of dollars) to which an investment P will grow if it is invested for n years and compounded annually at a rate of interest r. The value of a formula such as this lies in the fact that P, r, and n may change from one case to another, but A can always be determined.

However, a formula has a certain amount of mystery until the language of algebra is understood. This language consists of certain basic agreements we must make, along with the application of rather formal rules called axioms, which we assume to be true.

Many of the necessary agreements have already been discussed in previous chapters in arithmetic and should be reviewed at this time.

The symbols, +, −, ÷, will represent the familiar operations of addition, subtraction, and division. However we shall *discontinue* the use of × to represent multiplication since it would be confused with the script letter x. Thus, if w represents a number and if y represents a number, then the product of these two numbers will be represented by $w \cdot y$ or $(w)y$ or $(w)(y)$ in the same manner that multiplication was indicated earlier. However, the most common manner for indicating multiplication of two literal numbers (letters representing numbers) is the placement of the letters in juxtaposition, or one next to the other, as: wy.

Always consider wy as a *single* number because the product of two numbers is exactly one number. When w represents the number 7 and y represents the number 9 then wy means (7) (9), which is 63. The previous sentence is usually written:

If $w = 7$ and $y = 9$ then $wy = (7)(9) = 63$.

EXERCISES

Determine the required product in 1 – 5.

1. If $w = 3$ and $y = 8$ then $wy =$ __________

2. If $w = 2\frac{1}{2}$ and $y = 7\frac{1}{2}$ then $wy =$ __________

3. If $w = 2.16$ and $y = 14.2$ then $wy =$ __________

4. If $w = 2\frac{1}{2}$ and $y = .007$ then $wy =$ __________

5. If $w = 2$, $x = 3$, $y = 4$, and $z = 10$ then:

   (a) $wx =$ ________ (e) $ww =$ ________

   (b) $wz =$ ________ (f) $xy =$ ________

   (c) $xz =$ ________ (g) $wy =$ ________

   (d) $yx =$ ________ (h) $yy =$ ________

6. How else might yy be indicated?

7. Notice that the results of 5(d) and 5(f) are identical. What property of multiplication would seem to justify this? __________

8. What meaning should be attached to $(xy)z$? __________

9. If we assume xyz to be the same number as $(xy)z$ then when x is 3, y is 8, and $z = \frac{1}{4}$, $xyz =$ __________

10. Evaluate $(xy)(zw)$ if $x = 3$, $y = 0$, $z = 4$ and $w = 26$.

____

Multiplication is also indicated when a number is written in juxtaposition with a letter representing a number.

3y indicates the product of 3 and y. If y should be 2 then $3y = 3 \cdot 2 = 6$. y3 would indicate the same thing but 3y is preferred.

If a is a number and if b is a number then a + b represents the sum of a and b, a − b represents the difference of a and b; $\frac{a}{b}$ represents the quotient of a and b. Less frequently a ÷ b is used to indicate the quotient of a and b.

Parentheses are used in algebra, just as in earlier chapters in arithmetic, to group together numbers (or letters used to represent numbers) that are to be considered as a single number. They also indicate the *order* in which some operations are done. If a, b, and c are numbers then a − (b + c) represents the difference of two numbers. They are _____ and _____. The parentheses indicate that b and c must be added *before* the subtraction is done. In short, the parentheses indicate, "do this first."

Example: Evaluate $x + 2 - (x + y)$ if $x = 2$ and $y = 1$.

Solution:
$$\begin{aligned} x + 2 - (x + y) &= 2 + 2 - (2 + 1) \\ &= 4 - (3) \\ &= 1 \end{aligned}$$

EXERCISES

Evaluate the following is $x = 3$, $y = 2$, and $z = 4$.

1. $x - y + z$ ____________

2. $x - (y + z)$ ____________

3. $2(x + 1) - (z - y)$ ____________

4. $x + 2x - (z + y)$ ____________

5. $(x + 2) - (y + 2)$ ____________

TERMS, ALGEBRAIC EXPRESSIONS, AND FACTORS

Each number in an indicated sum is called a *term.* A + B indicates the sum of A and B and both of the numbers A and B are terms. The *sum* x + 2y + 3z has 3 terms. They are: ________, ________, and ________. Notice that 2y and 3z are single terms.

In arithmetic we have learned that 3 − 2 means 3 + (−2). This is also true in algebra, where a − b means a + (−b). Thus the terms in x − y are x and −y. The terms of 4x + (9 + y) − 6 are ________, (______), and ________.

EXERCISES

Identify the terms in the indicated sums.

	Number of terms	*List the terms*
Example: 2a + x + 6	3	2a, x, y
1. (a + b) + c	____	__________
2. 3a	____	__________
3. 5x + 2a + y	____	__________
4. 7(a + b) + c	____	__________
5. (a + b) + (c + d)	____	__________
6. 3(x + y) + 6	____	__________
7. 2x + 2y − 3	____	__________
8. x − y + 2A	____	__________
9. (x − y) + 2A	____	__________
10. 2A + x + 6	____	__________
11. 2A + (x + 6)	____	__________
12. (3A + 4) − (y + z)	____	__________

Algebraic expression is a general name given to any collection of numbers connected by algebraic operations. For example, 2x, a + b, 7, $\sqrt{3x}$, $2x^2 + 4x - 3$, are all algebraic expressions if a, b, and x represent numbers.

Another word we frequently meet in algebra is *factor.* It is used both as a noun and a verb. At this time we shall consider the noun. Each number in an indicated product is called a factor. Also any combination of these factors in product form is also a factor of the indicated product. The number 1 is a factor of every number, since each number a can be written as a · 1. We will omit this seldom used factor in the remainder of this section.

Thus, the factors of xy are x and y. The factors of 2ab are 2, a, b, 2a, 2b, and ab. Factors of 7(a + b) are 7 and (a + b). Note that there are no *terms* in this latter expression, other than the expression itself, since it is an indicated product rather than an indicated sum.

It is important to note also that no expression can contain both factors and terms at the same time. Much of algebra, however, involves the changing of the form of an expression from a sum of terms to a product of factors or visa versa. The manipulations will come later.

EXERCISES

Identify factors or terms in the following expressions.

	Factors	*Terms*
1. 3x + y	__________	__________
2. 3(x + y)	__________	__________
3. (5x − 3y) + 3	__________	__________
4. 2(x + y) + 5	__________	__________
5. 7x	__________	__________
6. 5x + 5y	__________	__________
7. 5(x + y)	__________	__________
8. (2x + 3) + (7x − 1)	__________	__________
9. (a + b) (c + d)	__________	__________
10. a + b(c + d)	__________	__________

Now that we have these agreements concerning the symbols used in algebra we will proceed to find the value of an algebraic expression when given the value of the letters involved.

Example: Evaluate $3x + 5y - 2z$, if $x = 1$, $y = 7$, $z = 3$

Solution:
$$3x + 5y - 2z = 3(1) + 5(7) - 2(3)$$
$$= 3 + 35 - 6$$
$$= 38 - 6$$
$$= 32$$

EXERCISES

Evaluate the following expressions if $x = 2$, $y = 3$, $z = 4$.

1. $x - y + z =$

2. $3(x + y) - z =$

3. $(x + y) - (x + z) =$

4. $x + y - (x + z) =$

5. $x + y - x - z =$

6. $xy + x =$

7. $5xy - 2yz =$

In exercises 8–12, $x = 8$, $y = 1$, and $z = 3$.

8. $x - y + z =$

9. $2(x + y) - z =$

10. $2x + y - z =$

11. $2(x - 4y) - 2z =$

12. $3(x + y) + 3z =$

In exercises 13–16, $x = \frac{1}{2}$, $y = \frac{1}{4}$, and $z = 2$.

13. $8(x - y) - z =$

14. $3x + 2y + 2z =$

15. $\frac{1}{2}x + 3y + z =$

16. $4(x + y) - z =$

FORMULAS

A *formula* is a statement that two algebraic expressions are equal. $F = \frac{9}{5}C + 32$ is a formula for changing Centigrade temperature to Fahrenheit. If we know the value of C we can find the value of F.

Example: Suppose C = 25; find F.

$$F = \frac{9}{5}C + 32$$

$$F = \frac{9}{5}(25) + 32$$

$$F = 45 + 32$$

$$F = 77$$

In other words 25° Centigrade is the same as 77° Fahrenheit.

It is interesting to note that we need not know what a formula means or how it was derived in order to use it, since it is simply a matter of evaluating an algebraic expression.

Example: $f = \frac{w}{2\pi}$ is a formula from electronics.

Find f if w = 88 and $\pi = \frac{22}{7}$.

$$f = \frac{w}{2\pi}$$

$$f = \frac{88}{2\left(\frac{22}{7}\right)}$$

$$f = \frac{88}{\frac{44}{7}}$$

$$f = 88\left(\frac{7}{44}\right)$$

$$f = 14$$

Note: The value of π is not exactly $\frac{22}{7}$ but this is a good approximation for most purposes.

EXERCISES

1. $I = Prt$ Find I if $P = 450$, $r = .06$ and $t = 3$.

2. $D = rt$ Find D if $r = 40$ and $t = 8\frac{1}{2}$.

3. $A = \frac{1}{2}bh$ Find A if $b = 7$ and $h = 10$.

4. $V = lwh$ Find V if $1 = 8$, $w = 3$ and $h = 4\frac{1}{2}$.

5. $A = \frac{1}{2}(a + b)h$ Find A if $a = 7$, $b = 9$ and $h = 3$.

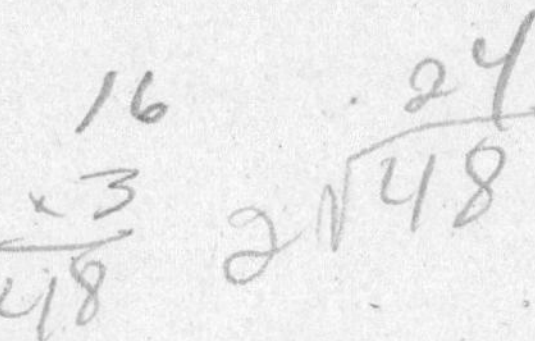

6. $C = 2\pi r$ Find C if $\pi = \frac{22}{7}$ and $r = 35$.

7. $V = \frac{1}{3}\pi r^2 h$ Find V if $r = 12$ and $h = 28$.

THE MEANING OF WHOLE NUMBERED EXPONENTS

In a previous chapter we studied the exponent, and you will remember, for example, that $4^3 = (4)(4)(4)$. In the same manner, $a^3 = (a)(a)(a)$; or in general, $a^n = (a)(a)(a)\ldots(a)$ where a is a factor n times. This definition applies only if n is a positive whole number. a^n is read "a to the nth power" or simply "a to the nth."

EXERCISES

Fill in the blanks.

1. $x^3 = (\_)(\_)(\_)$
2. Evaluate: x^4 if $x = 2$ _____
3. Evaluate: $x^3 + y^2$ if $x = 3$ and $y = 5$ _____
4. $x^2 \cdot y^3 =$ _____ if $x = 2$ and $y = 4$
5. $x^2 \cdot y^4 + y^3 =$ _____ if $x = 2$ and $y = 3$

CONVENTIONS ON THE ORDER OF OPERATIONS

What does the expression $2x^3$ mean?

There are two possibilities:
(1) if 2x is raised to the 3rd power then $2x^3$ is $(2x)(2x)(2x)$
(2) if only x is raised to the 3rd power then $2x^3$ is $(2)(x)(x)(x)$

Since these two possibilities are different it is obvious that we must agree on the one to be used. The question involved is "which operation, multiplication or raising to a power, will be performed first?" The answer we give to this question should conform to the generally accepted conventions on the order of operation. The conventions we will accept are listed here:

When an algebraic expression contains more than one operation, these operations will be performed in the following order.

1. raising to a power (exponents)
2. multiplication
3. division
4. addition and subtraction in the order in which they appear, reading from left to right.

As we have already stated, parentheses group together numbers that are to be considered as a single number. This means that parentheses will take precedence over all of the operations just listed.

Example: Evaluate $3 + 4x^3$ if $x = 2$.

Solution:
$$\begin{aligned} 3 + 4x^3 &= 3 + 4(2^3) \\ &= 3 + 4(8) \\ &= 3 + 32 \\ &= 35 \end{aligned}$$

Example: Evaluate $(3 + 4)x^3$ if $x = 2$

Solution:
$$\begin{aligned} (3 + 4)x^3 &= (3 + 4)\,(2^3) \\ &= 7(2^3) \\ &= 7(8) \\ &= 56 \end{aligned}$$

Compare the two examples. Notice the order of operations, and pay special attention to the use of exponents and the use of parentheses.

EXERCISES

Evaluate the following expressions. In problems 1 through 5 let $x = 3$, $y = 2$, and $z = 4$.

1. $5x^2 + y^3 =$

2. $7 + 6x^2 =$

3. $5x^2 + xy =$

4. $4 + 3z \div y - x =$

5. $3x^2 + 2y - z =$

In problems 6 – 11 let $a = 1$, $b = 2$, and $c = 3$.

6. $(a^2 + 4)^2 - 2c^2 =$

7. $(a^2 + 4)^2 - (2c)^2 =$

8. $5b^2 - 3a + 2c =$

9. $5b^2 - (3a + 2c) =$

10. $a^2 - b^2 + c^2 =$

11. $3a + a(b + c) =$

The agreements we have made so far have been concerned with the notation or language of algebra. These same agreements will also help us represent, in the symbolic form of an algebraic expression, many of the mathematical ideas found in sentences or phases.

Example: Write an algebraic expression representing the product of five and the sum of a and b.

Solution: $5(a + b)$

Example: Write an algebraic expression representing the sum of the product of five and a and the product of 3 and b.

Solution: $5a + 3b$

EXERCISES

Write an algebraic expression for each of the following.

1. The product of a and b ____________
2. The sum of a and (b + c) ____________
3. The product of a and (b + c) ____________
4. The fourth power of x ____________
5. The sum of a and the fourth power of x ____________
6. The fourth power of the sum of a and x ____________
7. A less than b ____________
8. The difference of x and y ____________
9. The square of y ____________
10. The square of the product of x and y ____________
11. The product of x and the square of y ____________
12. The quotient of a and b ____________
13. The quotient of x and (a + b) ____________
14. The sum of x and the quotient of a and b ____________
15. 7 diminished by x ____________
16. 3 more than y ____________

17. x more than the product of y and z __________

18. The cost of x articles at y dollars each __________

19. Twice the number $(a + 2b)$ __________

20. The sum of the squares of a and b __________

21. The square of the sum of a and b __________

22. The number of quarters equivalent to x dollars __________

23. Five more than x __________

24. The first counting number larger than the counting number y __________

25. The first even counting number larger than the even counting number x __________

26. The product of $(a + b)$ and $(c + d)$ __________

27. The cost of 5 articles at x dollars each and 3 articles at y dollars each __________

28. The amount of money collected when x one dollar tickets and y two dollar tickets are sold __________

29. Three times the sum of m and n __________

30. The product of a and the fourth power of y __________

31. The sum of x and 3 times y __________

32. The product of 7 and the sum of x and y __________

33. The sum of a and b subtracted from x __________

34. x less than the sum of a and b __________

35. The product of x and $(a + b)$ __________

36. The cost of 5 articles at $(x + y)$ cents each __________

The manipulations of algebraic expressions and symbols follow patterns that conform to certain basic rules, called *axioms.* The axioms necessary for our study follow.

Axiom 1. *The commutative axiom for addition.* The sum of any two numbers is independent of the order in which they are added:

$$a + b = b + a$$

Axiom 2. *The commutative axiom for multiplication.* The product of any two numbers is independent of the order in which they are multiplied:

$$ab = ba$$

Axiom 3. *The associative axiom for addition.* For any three numbers, a, b, and c, the number resulting from adding c to the sum of a and b is the same as the number resulting from adding the sum of b and c to a:

$$(a + b) + c = a + (b + c)$$

Axiom 4. *The associative axiom for multiplication.* For any three numbers, a, b, and c, the number resulting from multiplying the product of a and b by c is the same as the number resulting from multiplying a by the product of b and c:

$$(ab)c = a(bc)$$

The facts stated by these first four axioms are already familiar to us. We used these same properties in our work with arithmetic. We should note that these are axioms about the *operations* of addition and multiplication and *not* about the real numbers.

EXERCISES

1. Is subtraction commutative? Does $a - b = b - a$ for all numbers a and b?

   ______________ Give a counter example. ______________

2. Is division commutative? ______________

   Prove your answer. ______________

3. Does $a-(a - c) = (a - b) - c$ for all numbers a, b, and c?

   ____

4. Does $a \div (b \div c) = (a \div b) \div c$ for all numbers a, b, and c?

   ____

Many of the manipulations we wish to perform will be based on these four axioms. The following example illustrates how the axioms might be used to verify that a particular equation is valid.

Example: Prove that $(a + b) + c = (a + c) + b$ for all numbers a, b, and c.

At first glance this may look like an axiom. However, closer examination will show that the grouping is changed and that the order is also changed. To prove such a statement it is necessary to consider one side of the proposed equality and by careful use of the axioms try to change its form until it is identical to the other side. Study each step of this proof carefully.

Proof:

Statement	Axiom
1. $(a + b) + c = a + (b + c)$	1. associative for addition
2. $a + (b + c) = a + (c + b)$	2. commutative for addition applied to $(b + c)$ only
3. $a + (c + b) = (a + c) + b$	3. associative for addition

These three steps prove that $(a + b) + c = (a + c) + b$. Notice that step 1 changes $(a + b) + c$ into a second expression, which is then changed in step 2 to a third expression, and this expression is changed in step 3 to $(a + c) + b$. Hence, by using the axioms, we have shown that the left side of the proposed equality is equal to the right side and therefore both sides represent the same number.

A formal course in algebra will require many proofs such as the foregoing one, and others that will verify all manipulations of algebraic expressions.

EXERCISES

The following equalities can each be verified by one of the four axioms listed on page 259. Write the name of the proper axiom in the blank provided.

1. $a + (b + c) = (b + c) + a$ __________
2. $x(yz) = (xy)z$ __________
3. $(a + b)(c + d) = (c + d)(a + b)$ __________
4. $(a + b)(c + d) = (b + a)(c + d)$ __________
5. $(2x + y) + (3x + 4) = [(2x + y) + 3x] + 4$ __________
6. $(x + y)z = z(x + y)$ __________
7. $(x + y)z = (y + x)z$ __________
8. $(a + b) + c = a + (b + c)$ __________

We shall proceed now to list other axioms that are necessary for a study of algebra.

Axiom 5. *Axiom on zero.* For any number a, the number that results from adding a to zero is a:

$$a + 0 = a \quad \text{and} \quad 0 + a = a$$

Axiom 6. *Axiom on the negative of a number.* For each number a, there exists a number (−a), called *the negative of a,* such that the result of adding a to its negative is zero:

$$(-a) + a = 0 \quad \text{and} \quad a + (-a) = 0$$

Axiom 7. *The axiom on one.* For any number a, the number that results from multiplying one by a is a:

$$1 \cdot a = a \quad \text{and} \quad a \cdot 1 = a$$

Axiom 8. *The axiom on the reciprocal of a number.* For each number a (except 0) there exists a number $\frac{1}{a}$, called the reciprocal of a, such that the result of multiplying a by its reciprocal is 1:

$$\text{if } a \neq 0 \text{ then } \frac{1}{a} \cdot a = 1 \quad \text{and} \quad a \cdot \frac{1}{a} = 1$$

Axiom 9. *The distributive axiom.* For any three numbers a, b, and c, the number resulting from multiplying a by the sum (b + c) is the same as the number resulting from adding the two products (ab) and (ac):

$$a(b + c) = ab + ac$$

We noted that axioms 1 through 4 concerned operations. Notice that axioms 5 through 8 are axioms about specific numbers and their properties; these are referred to as *existence axioms.*

Axiom 9 is another axiom about operations. Notice that it involves both multiplication and addition.

The number zero is called the *additive identity element.* The set of real numbers has only one additive identity element. If $x + a = a$ then $x = 0$. If $(a + b + c) + x = x$ then $(a + b + c) = 0$.

The number 1 is called the *multiplicative identity element.* The set of real numbers has only one multiplicative identity element. If $x \cdot a = a$, then $x = 1$. If $(a + b + c) \cdot x = x$, then $(a + b + c) = 1$.

The uniqueness of the two identity elements is very important to the structure of algebra.

USING THE DISTRIBUTIVE AXIOM

No axiom is more important than another, but one of the most commonly used is the distributive axiom. A very large portion of elementary algebra is based on the application of this axiom. Let's take a closer look at its meaning.

Example: Evaluate $5(4+8)$ two different ways.

Solution: (a) $5(4+8) = 5(12) = 60$

(b) $5(4+8) = 5(4) + 5(8) = 20 + 40 = 60$

Some multiplication may be simplified by using the distributive axiom, especially if one wishes to multiply mentally.

Example: Multiply 12 by 33

Solution:

$$\begin{aligned}(12)(33) &= 12(30+3)\\ &= (12)(30) + (12)(3)\\ &= 360 + 36\\ &= 396\end{aligned}$$

EXERCISES

Evaluate each by (a) adding before multiplying, and (b) using the distributive axiom.

		(a)	(b)
1.	$9(12+8)$	__________	__________
2.	$14(2+7)$	__________	__________
3.	$4(8+6)$	__________	__________
4.	$7(9+10)$	__________	__________
5.	$8(4+11)$	__________	__________
6.	$12(7+8)$	__________	__________
7.	$5(10+7)$	__________	__________

A desirable extension of the distributive axiom is based on another axiom. Suppose we have an expression in the form of $(b+c)a$. Obviously, the distributive axiom does not apply directly. But $(b+c)a = a(b+c)$ by the commutative axiom for multiplication, and $a(b+c) = ab + ac$ by the distributive axiom. Therefore, $(b+c)a = ab + ac$.

Notice also that $ab = ba$ and $ac = ca$ by the commutative axiom for multiplication. Hence $(b+c)a = ba + ca$.

FACTORING

If $a = b$ then $b = a$ is called the symmetric property of equality. This is a very simple property, which we would generally use without regarding it as being important. In some instances, however, it gives us a different view of a statement.

When we write the distributive property as $a(b + c) = ab + ac$ we think of it as changing a product of factors into a sum of terms. However, if we write this property as $ab + ac = a(b + c)$ we think of it as changing a sum of terms to a product of factors. This second process is sometimes referred to as "undistributing," "removing the common factor," or simply "factoring."

Examples: Factor the following:

(a) $5x + 5y$

Solution: $5x + 5y = 5(x + y)$

(b) $3x + 12w$

Solution: $3x + 12w = 3x + 3 \cdot 4w = 3(x + 4w)$

(c) $6xy + 18x$

Solution: $6xy + 18x = 6xy + 6 \cdot 3 \cdot x = 6x(y + 3)$

Notice that you can always check factoring by using the distributive property.

EXERCISES

Factor each of the following expressions by removing the greatest common factor of the two terms.

1. $7x + 7y$ __________

2. $2a + 2b$ __________

3. $15x + 75y$ __________

4. $9ab + 9ac$ __________

5. $7x + 11xy$ ________________

6. $7a + 14b$ ________________

7. $2a + 2$ ________________

8. $9x + 27y$ ________________

9. $10x + 100xy$ ________________

10. $12ab + 84$ ________________

11. $15a + 3$ ________________

12. $7xy + 7x$ ________________

13. $3x^2y + x^2w$ ________________

14. $5x^2 + 10$ ________________

15. $13x^3y + 169x^2z$ ________________

16. $x^3 + x^2y = x^2 \cdot x + x^2y$ ________________

17. $3x^3 + 3x^2y$ ________________

18. $4.2x + 1.4y^2$ ________________

19. $28w + 196x$ ________________

20. $\sqrt{2}\, x^2 + \sqrt{2}\, y^2$ ________________

The factored form of an expression is not unique. For instance, $6x + 12$ can be written as $2(3x + 6)$, $3(2x + 4)$, $6(x + 2)$, $\frac{1}{2}(12x + 24)$, and so forth. However, the most useful form is generally the one in which the greatest common factor is removed and no new fractions are introduced.

An extension of the distributive axiom becomes apparent when we take a closer look at another axiom, the associative axiom for addition. Since $(a + b) + c = a + (b + c)$ we can agree to write $a + b + c$ without parentheses. In fact any expression of a sum of terms will be unchanged by the grouping of terms. For this reason we can factor an expression such as $3xy + 6x + 12ax + 9$ if we agree to extend the distributive axiom to multiplication over a sum of any number of terms. The expression in factored form is:

$$3(xy + 2x + 4ax + 3)$$

The distributive axiom is also used to show that: $a(b-c) = ab - ac$. Since $b - c$ means $b + (-c)$ it follows that: $a(b - c) = a(b + (-c)) = ab + a(-c) = ab + (-ac) = ab - ac$.

Thus, $3(a - b) = 3(a + (-b)) = 3a + 3(-b) = 3a + (-3b) = 3a - 3b$.

In practice this work is shortened to: $3(a - b) = 3a - 3b$.

Examples: Apply the distributive axiom to the following. Use the short method.

(a) $x(x - y^2) = x^2 - xy^2$

(b) $4a(3 - 6ab) = 12a - 24a^2b$

(c) $w(3x - 7) = 3wx - 7w$

In (c) could 3wx be written as 3xw? ______ Why? __________________

EXERCISES

Factor by removing the greatest common factor.

1. $2x + 6xy + 8ax$ ____________________
2. $9xy + x + 2xz$ ____________________
3. $7x + 35a + 14y + 42z$ ____________________
4. $x + ax + 3xy$ ____________________
5. $6ab + 4b + 2$ ____________________
6. $11xy + y$ ____________________
7. $7(x + y) + 2a(x + y)$ ____________________
8. $5(a + b) + c(a + b)$ ____________________

In the last two exercises of the previous set it was necessary to regard a sum as a single number and use it as a factor. When such sums are not already indicated, the factoring of such expressions may require more than one step.

Example: Factor $5ax + xb + 10a + 2b$

Solution: $5ax + xb + 10a + 2b$

$= x(5a + b) + 2(5a + b)$

$= (5a + b)(x + 2)$

Example: Factor $ax + ay + bx + by$

Solution: $ax + ay + bx + by$

$= a(x + y) + b(x + y)$

$= (x + y)(a + b)$

EXERCISES

Factor.

1. $2x + 2y + ax + ay$ ____________
2. $abx + aby + cx + cy$ ____________
3. $5ab + b + 10a + 2$ ____________
4. $35xy + 7x + 20ay + 4a$ ____________
5. $abcx + aby + cx + y$ ____________
6. $ab(x + y) + ab + 5x + 5y + 5$ ____________
7. $3x + 3y + ax + ay + x + y$ ____________
8. $xy + xz + 4y + 4z$ ____________
9. $144 + 2x + 72 + x$ ____________
10. $2x - 2y + ax - ay$ ____________

THE NEGATIVE OF A SUM

If a and b are numbers, the *negative* of the *sum* of a and b is written $-(a + b)$. Other forms of this expression are sometimes needed and these can be found by using the distributive property.

The number -3 has been shown to be the same as $(-1)(3)$. With literal numbers the same principle applies. That is,

$-a$ means $-1 \cdot a$,

$-xy$ means $-1 \cdot xy$ and

$-(a + b)$ is the same as $-1(a + b)$.

Follow this string of equations from beginning to end.

$$\begin{aligned} -(a + b) &= -1(a + b) \\ &= (-1 \cdot a) + (-1 \cdot b) \\ &= (-a) + (-b) \\ &= -a - b \end{aligned}$$

Can you show where the distributive property was used? Usually the steps are omitted and we write: $-(a + b) = -a - b$.

Some other examples are:

1. $-(a + 7) = -a - 7$
2. $-(a + 3c) = -a - 3c$
3. $-(xy + z + 8) =$ ______

When $-(a - b)$ is encountered, it can be written as $-a + b$, since

$$\begin{aligned} -(a - b) &= -1(a - b) \\ &= -1(a + (-b)) \\ &= (-1a) + ((-1)(-b)) \\ &= -a + (b) \\ &= -a + b \end{aligned}$$

Examples:

1. $-(a - 3) = -a + 3$
2. $-(4x + y) = -4x - y$
3. $-(a + b - c) = -a - b + c$

Example 3 uses both of the above rules.

We should note that $-a$ may not be a negative number. If $a = 0$ then $-a$ is also 0, and if a is less than 0 (written $a < 0$) then $-a$ is *positive.* For example, suppose $a = -4$; then $-a$ is $-(-4)$, which is 4.

Only when a is a positive number (written $a > 0$) will $-a$ be negative.

ADDING SIMILAR TERMS

We now wish to investigate the possibility of combining algebraic terms. We will find that some terms can be combined and some cannot. Our task is to discover the characteristics that determine which terms can be combined.

First let us consider the operation of addition that we have been using in arithmetic. Ask yourself the question, "What kinds of things have I added?" Your answer should be "real numbers," for we have applied this operation only to whole numbers, positive fractions, and directed numbers. In fact, as we have defined addition, *only real numbers can be added.*

If only real numbers can be added, how can we contemplate combining algebraic terms? It is only with the use of the distributive property that this is possible.

Example: Factor $2x + 5x$.

Solution: $2x + 5x = x(\_\_ + \_\_)$

Now if we have $2x + 5x$ in the form of $x(2 + 5)$ we see that the operation of addition can be performed on the numbers 2 and 5.

So $2x + 5x = x(2 + 5) = x(7)$ or $7x$

We have succeeded in combining the terms $2x$ and $5x$ and have shown that $2x + 5x = 7x$.

Can we combine $2ax$ and $5x$? We will attempt to use the same method as in the preceding example.

$$2ax + 5x = x(2a + 5)$$

Can we now add $2a$ and 5? ________. The answer is no, because we have no given number value for a.

Example: Combine $7xy$ and $9xy$.

Solution: $7xy + 9xy = (7 + 9)xy$

$= 16xy$

EXERCISES

Use the method of the preceding example to combine the terms in each problem. If the term cannot be combined write "not possible." (Remember that $1 \cdot A = A$, for any real number A.)

1. 4a and 3a ______
2. 2ax and 4b ______
3. 8x and x ______
4. 17xyz and 40xyz ______
5. 15x and −20x ______
6. 37xy and 3x ______
7. 14pq and 14pq ______
8. 44ac and −7y ______
9. 5(a + b) and 7(a + b) ______
10. 2(x − y) and 3(x − y) ______

Terms that have the same literal factors are called similar terms.

When you wrote "not possible" for Exercises 2, 6, and 8 in the preceding set of problems you were using the following rule:

"Algebraic terms can be combined only if they are similar."

By combining similar terms in an algebraic expression we can write the expression in fewer terms. Usually these operations are performed mentally because we can recognize similar terms without actually factoring.

EXERCISES

Combine the similar terms in each of the following. Do the operation mentally when possible and write only the answer.

1. $(7x + 2x) + 3x$ ________________
2. $5x - (3x + x)$ ________________
3. $(4x + 5x) - 7x$ ________________
4. $8a + 2a$ ________________
5. $9ax + 11ax + 3ax$ ________________
6. $5x + 3x + 2y$ ________________
7. $(2x + 3y) + 5x - y$ ________________
8. $(7ab + 2y) - 2ab$ ________________
9. $(5ay + 7) + (3ay - 4)$ ________________
10. $(3y - y) + (2x + 2y)$ ________________

In the last few exercises of the preceding set it was necessary for you to regroup and rearrange the terms before you could combine them. Probably you did these manipulations mentally. If so, you were making use of the commutative and associative axioms for addition. These axioms are so frequently used that we find it convenient to combine them in the following principles.

THE ASSOCIATIVE COMMUTATIVE PRINCIPLES

Addition:

For any collection of numbers, the sums obtained when these numbers are first arranged in any order (commuted), then grouped (associated) in any way, and then added are equal.

Multiplication:

For any collection of numbers, the products obtained when these numbers are first arranged in any order, then grouped in any way, and then multiplied are equal.

EXERCISES

Use the above principles to simplify the following expressions by combining similar terms.

1. $2x + y + x$ ____________________

2. $3ax + 4ax$ ____________________

3. $14x^2 + 3y + 2x^2$ ____________________

4. $\frac{1}{3}x + \frac{2}{5}y + \frac{3}{5}y$ ____________________

5. $.21x + .3y + .4x + .07y$ ____________________

6. $.09x^2y + 3.4z + z + 11.6$ ____________________

7. $3 + x^2 + 8 - x^2 - 8$ ____________________

8. $3x^2y^2 + x^2y^2 + 16 + 5x^2y^2$ ____________________

9. $11ab + c + 2ab + (-c)$ ____________________

10. $.4bc + .3cb + .3bc$ ____________________

11. $3a + b + 7b - \frac{1}{2}a$ ________________

12. $xy + 3xy + h + 3h$ ________________

13. $abc + cab + bca$ ________________

14. $x + y + xy + 4x$ ________________

15. $36x + 9x^2 + 4x^3 + (-9x^2)$ ________________

16. $36x + 9x^2 + 4x^3 - 9x^2$ ________________

17. $(a + b) + (2a + b)$ ________________

18. $(a + b) - (2a + 4b)$ ________________

19. $x^2 - (3x^2 + y) + 2y$ ________________

20. $ax + b + (3ax - 7b) + z$ ________________

21. $5ax + 2xy + 3x + 4ax - xy + x$ ________________

22. $16x^2 + 3x + 5x^2 - 2x$ ________________

23. $2ax^2 + 7y - 2y + 3ax^2$ ____________________

24. $14x^2y + 7ax + 11x^2y + x$ ____________________

25. $ax + by + 3ax + 5by$ ____________________

26. $4(a + b) + 3a - 2b$ ____________________

27. $3y - 4x + 6x - 2y$ ____________________

28. $x(x + y) + 3x^2 + 2xy$ ____________________

29. $4x(a + b) + 7a(x + y)$ ____________________

30. $7(2a + b) - 5a - 5b$ ____________________

31. $.2(.4x + 3) - .7x + .4$ ____________________

32. $7\frac{1}{2}x + 3(x + \frac{3}{5}) + 1\frac{1}{5}$ ____________________

33. $4(a + b) + 4(-a - b)$ ____________________

34. $.002x + 3.60 - .4(.08x - 1.36)$ ____________________

EQUATIONS

In an early chapter we discussed the ideas of number sentences and open sentences. These same concepts are used in algebra and the expressions for them are generally referred to as equations. You will recognize the similarities of these ideas in the following definitions.

An Algebraic Equation: A statement in symbols that two number expressions are equal is an algebraic equation.

Algebraic equations are of two types.

(a) *Identities.* An identity is an algebraic equation that is true for all possible replacements of the literal numbers. (Literal numbers in equations are generally referred to as *variables.*)

Example: $2x + 3x = x + 4x$ is an identity because any real number replacing x will result in a true number sentence.

(b) *Conditional Equations.* A conditional equation is true only for certain replacements of the variables.

Example: $x + 7 = 10$ is true for $x = 3$ but not for any other number.

Example: $x + y = 10$ is true if $x = 3$ and $y = 7$ and for many other values of x and y, but is false for some values of x and y.

Any replacement of variables that makes an equation a true number sentence is called a *solution* of the equation.

EXERCISES

Indicate which of the following are identities and which are conditional equations. Find at least one solution for each conditional equation by giving replacements for the variables that will result in a true number sentence.

1. $x = x$ ____________________

2. $2x = 1$ ____________________

3. $x + y = y + x$ ____________________

4. $x + 3 = 10$ ____________________

5. $x + y + x = 2x + y$ ____________________

6. $\frac{1}{2}y + \frac{3}{4}y = \frac{5}{4}y$ ____________________

7. $a(b + c) = ab + ac$ ____________________

8. $3x + 2 = 11$ ____________________

9. $x + y = 7$ ____________________

10. $3 + (x + y) = (3 + x) + y$ ____________________

11. $4a = 24$ ____________________

12. $16x + 3x = 76$ ____________________

13. $2y = y$ ____________________

14. $.3d + .8d + .40d = 3.0$ ____________________

15. $x^2 = 128$ ____________________
(answer correct to hundredths)

EQUIVALENT EQUATIONS

Two equations are said to be *equivalent* if every solution of one is also a solution of the other.

Example: $3x = 6$ and $4x = 8$ are equivalent, since replacing x by 2 will make a true sentence and this is the only solution for each of them.

EXERCISES

Which of the following pairs of equations are equivalent? Write "yes" for those that are equivalent and "no" for those not equivalent.

1. $2x + 7 = 11$ and $x = 2$ ________
2. $7x + 4 = 12x - 1$ and $x = 1$ ________
3. $5x - 3x = x + 2$ and $x = 4$ ________
4. $8x + 3 - 2x + 6 = 32$ and $x = 5$ ________
5. $3x - 2 - 4x + 10 = 6$ and $x = 2$ ________
6. $x - 1 = x + 1$ and $x = 0$ ________
7. $3x - 12 = x - 4$ and $x = 4$ ________
8. $2x + 3 - x + 1 - 4x = 3x$ and $x = 2$ ________

An equation like $x = 3$ has an obvious solution. The only possible solution is 3. In fact, any equation such as $x = k$, where k is a number, has k as a solution.

If we can take a more complicated equation and simplify it to an equivalent equation of the form $x = k$, then k will be the desired solution of the original equation.

For instance, suppose we wish to solve the equation $4x + 2 = 3x + 5$ and can find ways of simplifying it to the equivalent equation $x = 3$. Then we can say that 3 is a solution of $4x + 2 = 3x + 5$.

The key word in the above statement is *equivalent.* We must have certain rules that will allow us to find an equation equivalent to a given equation. These rules, together with the other manipulations we have already been performing, will enable us to solve many equations.

The following rules are stated as properties of equality:

(a) *Equality and addition.* If the same number expression is added to both sides of an equation, an equivalent equation results.

Examples: $3x + 1 = 7$ and $(3x + 1) + 5 = (7) + 5$ are equivalent.

$2x + 4 = 9$ and $(2x + 4) + x = 9 +$ ____ are equivalent.

$x - ½ = z$ and $(x - ½) + ½ =$ ______ are equivalent.

(b) *Equality and subtraction.* If the same number expression is subtracted from both sides of an equation, an equivalent equation results.

Examples: $x + 6 = 4$ and $(x + 6) - 1 = 4 - 1$ are equivalent.

$3x - 7 = 2x$ and $(3x - 7) - 2x = 2x -$ ______ are equivalent.

$3x + 2 = 7x + 1$ and $(3x + 2) - w =$ ______ are equivalent.

(c) *Equality and multiplication.* If both sides of an equation are multiplied by the same number expression (other than zero), an equivalent equation results.

Examples: $7x + 2 = 16$ and $3(7x + 2) = 3(16)$ are equivalent.

$4x - 1 = 7$ and $x(4x - 1) = 7($___$)$ are equivalent if ___ is not 0.

$¼x = 7$ and $4(¼x) =$ ________________ are equivalent.

(d) *Equality and division.* If both sides of an equation are divided by the same number expression (other than zero), an equivalent equation results.

Examples: $3x + 6 = 1$ and $\frac{3x + 6}{4} = \frac{1}{4}$ are equivalent.

$5x + 1 = 17$ and $\frac{5x + 1}{3x + 1} = \frac{17}{3x + 1}$ are equivalent if ____ is not zero.

$2x + 3 = 20$ and $\frac{2x + 3}{4} =$ ______ are equivalent.

Now let's see how these rules can be used to find the solution of an equation.

Example: Solve $4x + 3 = 11$

1.	$4x + 3 = 11$	
2.	$(4x + 3) - 3 = 11 - 3$	rule (b)
3.	$4x + (3 - 3) = 11 - 3$	associative axiom for addition
4.	$4x + 0 = 8$	$3 + (-3) = 0$. Axiom on the negative of a number.
5.	$4x = 8$	Axiom on 0
6.	$\frac{4x}{4} = \frac{8}{4}$	rule (d)
7.	$x = 2$	Since $\frac{4}{4} = 1$ and $\frac{8}{4} = 2$.

Because each equation is equivalent to the preceding one, $x = 2$ is equivalent to the original equation. 2 is the solution of $x = 2$ and therefore 2 is the solution of $4x + 3 = 11$.

A word of warning must be given here. Since any one of us might make a mistake in arithmetic, we need to check our solution. Is $x = 2$ really equivalent to $4x + 3 = 11$? The only way to know for sure is to replace x with 2 and see if a true number sentence results.

Check:

$$4x + 3 = 11$$
$$4(2) + 3 = 11$$
$$8 + 3 = 11$$
$$11 = 11$$

Since $11 = 11$ is a true number sentence, our solution is correct. In checking we must always substitute our supposed value for the variable in the *original equation.* If we substitute in any step other than the original equation, we could not know for sure that we have a solution. Why? ____

____

The above example is done in a more painstaking manner than might be necessary. You may wish to shorten this work by doing some steps mentally.

EXERCISES

Solve the following equations and check to verify your solution.

1. $2x + 10 = 40$ *Check:*

2. $3 + 7x = 52$ *Check:*

3. $.8w + .3 = 8.3$ *Check:*

4. $17x + 27 = 180$ *Check:*

5. $4z + 2 = 4$ *Check:*

6. $24x + 2 = 50$ *Check:*

7. $14x - 20 = 8$ *Check:*

8. $3w - 4 = 17$ *Check:*

9. $2.5 + x = 2.5$ *Check:*

10. $9x - 2\frac{2}{5} = \frac{3}{5}$ *Check:*

11. $8x + 2 = 42$ *Check:*

12. $13x - 13 = 156$ *Check:*

13. $4w + 7 = 75$ *Check:*

14. $w - \frac{3}{4} = 7\frac{1}{4}$ *Check:*

15. $3y + 3 = 4$ *Check:*

Example: Solve $5x + 4 = 2x + 13$

Solution:

$5x + 4 = 2x + 13$

$(5x + 4) - 4 = (2x + 13) - 4$ by rule ____________

$5x = 2x + 9$ by ____________

$5x - 2x = (2x + 9) - 2x$ by rule ____________

$3x = 9$ by ____________

$\frac{3x}{3} = \frac{9}{3}$ by rule ____________

$x = 3$ by ____________

Check:

$5x + 4 = 2x + 13$

$5(3) + 4 = 2(3) + 13$

$15 + 4 = 6 + 13$

$19 = 19$

EXERCISES

Solve and check.

1. $2x + 6 = x + 7$ *Check:*

2. $3x - 4 = x$ *Check:*

3. $9x - 4 = 5x + 4$ *Check:*

4. $2x + \frac{1}{2} = x + 4\frac{1}{2}$ *Check:*

5. $7x - 3 = 4x + 18$ *Check:*

6. $4x + 6 = 16 - x$ *Check:*

7. $3x - 2 = 14 - x$ *Check:*

8. $3(x + 1) = 2x + 7$ *Check:*

9. $10x = 2(x + 2)$ *Check:*

10. $4x + 11 = x - 3$ *Check:*

Example: $2(x + 6) + 3(x - 1) = x + 17$

Solution:
$$2(x + 6) + 3(x - 1) = x + 17$$
$$2x + 12 + 3x - 3 = x + 17$$
$$5x + 9 = x + 17$$
$$5x + 9 - 9 = x + 17 - 9$$
$$5x = x + 8$$
$$5x - x = x + 8 - x$$
$$4x = 8$$
$$x = 2$$

Check:
$$2(x + 6) + 3(x - 1) = x + 17$$
$$2(2 + 6) + 3(2 - 1) = 2 + 17$$
$$2(8) + 3(1) = 19$$
$$16 + 3 = 19$$
$$19 = 19$$

EXERCISES

Solve and check.

1. $3(x - 4) + 2(x + 6) = 5x$ *Check:*

2. $x + 8 + 3(x - 4) = 6 - x$ *Check:*

3. $5(x + 1) + 3(x - 1) = 3x + 12$ *Check:*

4. $6(x + \frac{1}{2}) + 2(x + 1) = 4x + 7$ *Check:*

5. $8x + 2(2x + 1) = 8$ *Check:*

6. $7(x + 1) = 2(x - 4) + x + 19$ *Check:*

7. $\frac{1}{2}(x + 6) + \frac{3}{2}x = 15 - 4x$ *Check:*

8. $\frac{3}{4}(x + 8) + 2(x - 1) = \frac{1}{4}(64 - x)$ *Check:*

9. $.6(x - 2) + x = .8 - .4x$ *Check:*

10. $8(x + 4) + 9(x - 3) = 15x + 6$ *Check:*

In Chapter 13 we discussed operations with directed numbers and developed rules for these. You should review those rules at this time. Study each example below very carefully, and pay special attention to the signs involved.

Example: Solve $2x + 7 = 3$

Solution:

$$2x + 7 = 3$$
$$2x + 7 - 7 = 3 - 7$$
$$2x = -4$$
$$x = -2$$

Check:

$$2x + 7 = 3$$
$$2(-2) + 7 = 3$$
$$-4 + 7 = 3$$
$$3 = 3$$

Example: Solve $5 - 2x + 3 = x + 17$

Solution:

$$8 - 2x = x + 17$$
$$8 - 2x - 8 = x + 17 - 8$$
$$-2x = x + 9$$
$$-2x - x = x + 9 - x$$
$$-3x = 9$$
$$x = -3$$

Check:

$$5 - 2x + 3 = x + 17$$
$$5 - 2(-3) + 3 = (-3) + 17$$
$$5 + 6 + 3 = 14$$
$$14 = 14$$

EXERCISES

Solve and check.

1. $5x + 20 = 5$ *Check:*

2. $2w + 7 = 3$ *Check:*

3. $x + 2 + 4x = -3$ *Check:*

4. $-3x = 12 + x$ *Check:*

5. $2(x + 7) = 20$ *Check:*

6. $5(x + 1) - 3 = -13$ *Check:*

7. $-x + 7 = 4$ *Check:*

8. $3y + 2(y + 10) = 0$ *Check:*

9. $2x + 15 = 4 + 13x$ *Check:*

10. $2(x + 3) = 5(x + 12)$ *Check:*

11. $3 + \frac{1}{2}(4x + 8) = -1$ *Check:*

12. $w + 8 - 7w = w - 6$ *Check:*

13. $4(x + 1) + 3(2x + 3) = 13$ *Check:*

14. $-12z - 7 + 4(z + \frac{1}{2}) = -1$ *Check:*

Example: Solve $2(x - 3) - 5(x + 4) = 3(x + 1) + 1$

Solution:

$$
\begin{aligned}
2(x - 3) - 5(x + 4) &= 3(x + 1) + 1 \\
2x - 6 - 5x - 20 &= 3x + 3 + 1 \\
-3x - 26 &= 3x + 4 \\
-3x - 26 + 26 &= 3x + 4 + 26 \\
-3x &= 3x + 30 \\
-3x - 3x &= 3x + 30 - 3x \\
-6x &= 30 \\
x &= -5
\end{aligned}
$$

Check:

$$
\begin{aligned}
2(x - 3) - 5(x + 4) &= 3(x + 1) + 1 \\
2(-5 - 3) - 5(-5 + 4) &= 3(-5 + 1) + 1 \\
2(-8) - 5(-1) &= 3(-4) + 1 \\
-16 + 5 &= -12 + 1 \\
-11 &= -11
\end{aligned}
$$

EXERCISES

Solve and check.

1. $4x - 2(x + 3) = (6 - 2x) - 10$ *Check:*

2. $3(x - 1) + 4x = 9 - 2(x - 1)$ *Check:*

3. $2x + 6 = 4(3 - x) + 6$ *Check:*

4. $x - 3(2 - x) = 4 - (x + 4) + 9$ *Check:*

5. $17x = 3(4x - 1) - 5(2x + 7) + 8$ *Check:*

6. $106 - 4(x + 12) = 34$ *Check:*

7. $10x + 3 = 5 - 2(x + 1)$ *Check:*

8. $9x + 11 - 3(x - 3) = 2$ *Check:*

9. $x - 7 = 7 - (x + 4)$ *Check:*

10. $3x + 2(x - 1) = 10 - x$ *Check:*

When the solution of an equation is a fraction or mixed number, checking can be more of a task.

Example: Solve $3(x + 4) - 7 = x + 6$

Solution:

$$3(x + 4) - 7 = x + 6$$
$$3x + 12 - 7 = x + 6$$
$$3x + 5 = x + 6$$
$$3x + 5 - 5 = x + 6 - 5$$
$$3x = x + 1$$
$$3x - x = x + 1 - x$$
$$2x = 1$$
$$x = \tfrac{1}{2}$$

Check:

$$3(x + 4) - 7 = x + 6$$
$$3(\tfrac{1}{2} + 4) - 7 = \tfrac{1}{2} + 6$$
$$3(4\tfrac{1}{2}) - 7 = 6\tfrac{1}{2}$$
$$13\tfrac{1}{2} - 7 = 6\tfrac{1}{2}$$
$$6\tfrac{1}{2} = 6\tfrac{1}{2}$$

EXERCISES

Solve and check.

1. $7x + 12 = 2x - 3$ *Check:*

2. $4x + 6 - 2(x + 1) = 7$ *Check:*

3. $5(x - 6) - 2(x - 2) = 6(x + 1)$ *Check:*

4. $12x - 3(x + 6) = 41$ *Check:*

5. $7\frac{1}{2}x + 9 = \frac{1}{2}x + 3(x + 1)$ *Check:*

6. $4(x - 1) + 6x = 9(x - 3) + 12$ *Check:*

7. $8(x + 6) = 14(x - 3) - 8x$ *Check:*

8. $3(x + 1) + 2(x + 1) - 6(x + 1) = 0$ *Check:*

9. $7x + 3 = 4 - 2(x + 6)$ *Check:*

10. $12 + 3x - 7 = 4(x - 6) + 11$ *Check:*

11. $2(x + 6) - 3(x + 4) = 14 - 5x$ *Check:*

12. $3.1x + 7 = 2.2x + .7$ *Check:*

13. $4(x + 2) - 3(2x - 1) = 7 - 2(4 + 3x)$ *Check:*

14. $9x - 36 = x + 2$ *Check:*

15. $9(x - 4) = 3(x + 2)$ *Check:*

16. $3(4 + x) - 7 = 4x - 13$ *Check:*

17. $5x = 3(x - 1) - 2(5x + 8) + 12$ *Check:*

18. $6(3 - x) + 7 = 12 - 5(14 - x)$ *Check:*

19. $3(x + 4) - 2(2x + 1) - 6(x - 7) = 112$ *Check:*

20. $x + 2(x + 1) - 3(x + 2) = 15$ *Check:*

21. $2x = 5(3 - x) - 1$ *Check:*

LITERAL EQUATIONS

It was stated previously that formulas could be used only if we became familiar with the language of algebra. We have now developed this language well enough to use many formulas from business and the sciences.

The formula $I = Prt$ is used in business. I is the amount of interest on P dollars at rate r % per unit time for t units of time.

Example: How much is the interest on \$500 for 3 years at 6% per year?

Solution:
$$I = Prt$$
$$I = (500)(.06)(3)$$
$$I = \$90$$

We used formulas like this one before and found that it only requires a little knowledge of the language of algebra. Let us now change the problem a little to see how our ability to solve equations can be useful.

Example: If the interest on a certain amount of money invested at 6% per year for 5 years is \$54.00, how much money is invested?

Solution: In the problem P is the unknown.
$$I = Prt$$
$$54 = P(.06)(5)$$
$$54 = .3\,P$$
$$P = \frac{54}{.3}$$
$$P = \$180$$

Example: If the interest on \$600 invested at 6% per year was \$21.00, for how long was the money invested?

Solution:
$$I = Prt$$
$$21 = (600)(.06)\,t$$
$$21 = 36\,t$$
$$\frac{21}{36} = t$$
$$t = \frac{7}{12} \text{ years or 7 months}$$

We can use $I = Prt$ and find the value represented by any letter if we know the value of each of the remaining letters. This is true because of our ability to solve equations. The following exercises will give practice with this particular formula.

EXERCISES

Use $I = Prt$ to solve each of the following.

1. If $I = \$12$, $P = \$400$, $t = 2$, find r.

2. If $P = \$1000$, $r = 3\%$ each 6 months, $t = 18$ months, find I.

3. If $I = \$15$, $P = \$50$, $r = 2\%$ per period, find t (the number of periods).

4. If $I = \$9$, $P = \$150$, $t = 2$ years, find r (the rate per year).

In the previous set of exercises we used the formula $I = Prt$ to find each of the four letters involved. It is not necessary to use numbers to solve an equation for a particular letter.

Example: Given $I = Prt$, express P in terms of I, r, and t.

Solution: $I = Prt$

$\frac{I}{rt} = \frac{Prt}{rt}$ dividing by rt

$P = \frac{I}{rt}$ Since $\frac{rt}{rt} = 1$ and $P(1) = P$.

Example: Solve $A = \frac{1}{2}bh$ for b in terms of A and h.

Solution: $A = \frac{1}{2}bh$

$2(A) = (\frac{1}{2}bh)(2)$ Multiply by 2

$2A = bh$ Since $(\frac{1}{2})(2) = 1$

$\frac{2A}{h} = \frac{bh}{h}$ Divide by h

$b =$ __________ Why? __________

Equations or formulas such as the foregoing are called literal equations. It is important to be able to solve a literal equation for any letter. We have enough rules for solving equations to solve for any letter in a literal equation if only the operations of addition, subtraction, multiplication, and division are used.

EXERCISES

Solve as directed.

1. Solve $I = Prt$ for t.

2. $V = r^2h$ is a formula for the volume of a circular cylinder. Solve for h.

3. $P = 2l + 2w$ gives the perimeter of a parallelogram in terms of the length and width.

   (a) Solve for l. ________________

   (b) Solve for w. ________________

4. $V = lwh$ is the formula for the volume of a parallelepiped. Find w in terms of V, l, and h.

5. $f = \frac{mv^2}{r}$ is a formula for centripetal force f exerted on a body of mass m traveling along a curve of radius r at velocity v. Solve for m.

6. $F = \frac{9}{5}C + 32$ is a formula for changing Fahrenheit to Centigrade measure of temperature. Solve for C.

7. The General Gas Law is expressed as $\frac{P_1V_1}{T_1} = \frac{P_2V_2}{T_2}$ where P, V, and T represent pressure, volume and temperature.
(Note: P_1 is read P-sub one and is not the same as P-sub two. For convenience the subscripts are used in many formulas instead of using a second letter. They must always be treated as being different literal numbers.)

(a) Solve for P_2.

(b) Solve for T_2.

8. Uniform motion is expressed by the formula $D = rt$. (Distance equals rate multiplied by the time.)

(a) Solve for r.

(b) Solve for t.

ANSWERS TO SELECTED PROBLEMS

Chapter 1, Page 3

1. Two hundred three
3. Two hundred three thousand
5. Thirty thousand sixty
7. One hundred thousand one hundred
9. One billion five hundred million
11. Two billion eight hundred forty six million three hundred seventy thousand

Page 4

1. Five thousand two hundred eighty
3. Seven thousand one
5. One hundred thirty four thousand one
7. 5,040
9. 431,000,000
11. 800,800,808
13. 7,943,532

Chapter 2, Page 9

1. 133
3. 1670
5. 1666
7. 18298
9. 1945
11. 8350

Page 12

1. 8
3. 17
5. 9
7. 6

Page 14

1. 113
3. 72
5. 1878

Page 14 cont

7. 358
9. 36

Page 16

1. 129
3. 2438
5. 55
7. 11059
9. 1703
11. 55
13. 8657
15. 47469
17. 411012
19. 6,640,030

Chapter 3, Page 19

1. 4
3. 17
5. 119
7. 1879
9. 6
11. 203
13. 72

Page 20

1. (a) Seven thousand nine hundred sixteen.
   (b) Ninety thousand ninety
2. 24,135
3. (a) 20,000 (b) 86435
4. (a) 88666 (b) 6516 (c) 1760
5. (a) $x = 22$ (b) 53 (c) 24
   (d) 1679

Page 22

1. 8 − (3 + 5) = 0
3. (15 + 6) − 5 = 16
5. 141 − [8 − (3 + 2)] = 138
7. 12 − [(8 − 3) + 5] = 2
9. (32 + 7) − 4 = 35

Page 24

1. (a) x = 9 (b) x = 9 (c) x = 3
3. (a) x = 3 (b) x = 3 (c) x = 17
5. (a) x = 0 (b) x = 0 (c) x = 20
7. 63
9. 104

Page 25

1. (27 + 3) + 16 = 46
3. (17 + 8) + 5 = 30

Page 26

1. Commutative
3. Associative
5. (17 + 3) + 7 = 27
7. (9862 + 8) + 4000 = 13870
9. (1600 + 400) + 210 = 2210

Page 27

1. 20 − 8
3. 9 + 7
5. 46 − (2 + 7)

Page 28

1. x + 5
3. 15 − x
5. x + 16
7. Y = x + 16
9. (2 + 7) − 5
11. 80 = x − 10
13. (2 + 12) + x

Page 30

1. $96
3. 24

Page 30 cont

5. (a) 1,168 ft. (c) Yes, 3,606,086 sq miles
7. 57
9. (a) 70,686,091 (c) 1850 to 1950

Page 33

1. (a) 22
2. (a) 15,1 (f) 12,20
4. (a) 14,17,20
   (c) 13,21,34
   (e) 21,17,26
6. 70 inches

Chapter 4, Page 37

1. (A + B) + C = A + (B + C)
3. A + B = B + A
5. Commutative property for multiplication
7. Associative property for multiplication

Page 38

1. (a) 22 × (20 + 3) = 22 × 23 = 506
   (b) (22 × 20) + (22 × 3) = 440 + 66 = 506
3. (a) 112 × (6 + 4) = 112 × 10 = 1120
   (b) (112 × 6) + (112 × 4) = 672 + 448 = 1120

Page 39

1. (a) 20 = 20
   (b) Associative property for addition
3. (a) 140 = 140
   (b) Commutative property for multiplication
5. (a) 34 = 34
   (b) Commutative property for addition

Page 41

1. 8004
3. 21346
5. 34132
7. 131,793

Page 41 cont

9. 42300
11. 872682

Page 43

1. 118958
3. 4,015,000
5. 9,984,636
7. 824,464
9. 44,068,320
11. 1,738,080
13. 918
15. 11736 lbs.
17. (b) 45 cu. ft.

Page 47

1. 83
3. 98
5. 1875
7. 43
9. 796

Chapter 5, Page 52

1. 5,10,20
3. 1,2,3,6
5. 1,5
7. 1,3,7,21
9. 1,2,19,38
11. 1,17
13. 1,5,7,35
15. 1,2
17. 1,2,4,5,8,10,20,25,40,50,100,200
19. 1

Page 53

3. This number has more than 2 divisors.
5. This number has more than 2 divisors.
7. This number has only one divisor.
9. This number is not a counting number.

Page 57

Answers will vary. Any numbers whose product is the given number will be correct for

Page 57 cont

this exercise.

Page 59

1. T	12. T
6. T	15. T
10. F	20. F

Page 60

1. $9 \cdot \underline{60}$
   $= 3 \cdot \underline{3} \cdot 12 \cdot \underline{5}$
   $= 3 \cdot 3 \cdot 3 \cdot 4 \cdot 5$
   $= 3 \cdot 3 \cdot 3 \cdot 2 \cdot 2 \cdot 5$
3. $36 = 6 \cdot 6$
   $= 2 \cdot 3 \cdot 2 \cdot 3$
   $= 2^2 \cdot 3^2$
5. $12 = 4 \cdot 3$
   $= 2 \cdot 2 \cdot 3$
   $= 2^2 \cdot 3$

Page 63

1. $2 \cdot 3^2$
3. $2^2 \cdot 3 \cdot 5$
5. $2 \cdot 61$
7. $2^3 \cdot 17$
9. $2^3 \cdot 5 \cdot 11$
11. $2^3 \cdot 7 \cdot 11$

Page 67

1. F.	11. 4
3. T	13. 4
5. F.	15. 12
7. F.	17. 13
9. T	

Chapter 6, Page 71

1. 12/15
3. 10/5
5. 8/10
7. 4/14
9. 13/5

Page 71 cont

11. 5/5
13. 10/8
15. 6/5
17. 3/4
19. 3/4
21. 32/32
23. 18/9

Page 72

1. Two million three hundred eighty four thousand sixty one
3. 5586
5. 1,827,783
7. (a) $2^4 \cdot 3^2$
9. 7/8

Chapter 7, Page 75

1. $\dfrac{2 \times 2 \times 2}{2 \times 2 \times 3} = \dfrac{2}{3}$
3. 3/8
5. 1/3
7. 69/91
9. 21/29
11. 1/26
13. 21
15. $2 \times 3 \times 7$
17. 54
19. 140
21. 4,1

Page 78

1. 47
3. 33
5. 105
7. 49
9. 123

Page 80

1. 3 1/2
3. 5 2/3
5. 11 2/10 or 11 1/5
7. 7 1/9

Page 80 cont

9. 7 1/2

Chapter 8, Page 82

1. 25
3. 14
5. 144
7. 3 3/4
9. 16

Page 84

1. 1/4
3. 7/20
5. 1/35
7. 63/128
9. 32/57

Page 86

1. 24/77
3. 2/7
5. 4
7. 4
9. 5
11. 145/16
13. 6/5
15. 1
17. $16
19. 287

Page 88

1. 3/4
3. 27
5. 1875/4 or 468 3/4
7. 7
9. 13048/3 or 4349 1/3

Chapter 9, Pages 91–94

1. 9
3. 36
5. 24/7 or 3 3/7
7. 27/7 or 3 6/7
9. 49/3 or 16 1/3

Pages 91–94 cont

11. 162
13. 64/11 or 5 9/11
15. 2
17. 18/25

Pages 95–97

1. 21/4 or 5 1/4
3. 15/8 or 1 7/8
5. 57/16 or 3 9/16
7. 196/9 or 18 4/9
9. 1

Pages 101–102

1. 1/12
3. 4/7
5. 35/156
7. 9
9. 22

Chapter 10, Page 104

1. 12,24,36,48,60, . . .
3. 7,14,21,28,35, . . .
5. 2041,4082,6123, . . .
7. 3/4, 6/4, 9/4, 12/4, 15/4, . . .
9. 7,14,21,28,35,42, . . .
11. 3,6,9,12,15,18,21,24,27,30,33, . . .

Page 105

1. 18
3. 44
5. 33
7. 78
9. 21

Chapter 11, Page 109

1. 6/3
3. 14/5
5. 12/12

Page 112

1. 8

Page 112 cont

3. 42
5. 12
7. 24

Pages 113–117

1. 19/24
3. 11/32
5. 153/30 or 5 1/10
7. 35/48
9. 36/24 or 1 1/2
11. 85/48 or 1 37/48
13. 7/48
15. 39/32 or 1 7/32
17. 137/60 or 2 17/60

Pages 119–121

1. 8 7/8
3. 8 5/8
5. 300
7. 18 1/4
9. 12 13/24

Pages 123–126

1. 1 5/8
3. 1 5/6
5. 6 5/9
7. 3 15/16
9. 3 2/3
11. 114 4/5
13. 2 7/12

Page 127

1. 16 29/80
3. 117/8 or 14 5/8
5. 8 41/48
7. 161/12 or 13 5/12
9. 21/40

Chapter 12, Page 130

1. 6/10
3. 3/100
5. 143/1000

Page 130 cont

7. 14/10000
9. 7550/10000

Page 131

3. five tenths
5. fourteen thousandths
7. eight hundred twenty two thousandths
9. nine thousand two hundred thirty four ten thousandths.

Page 135

1. 60.96
3. 14.003
5. 103.97
7. 47.44
9. 10.5533
11. 38.2212
13. 862.43
15. 15.165

Page 137

1. 42.26
3. 2.41759
5. 20.911
7. 11.464
9. 13.858
11. $10.81
13. $9.86
15. 7173.69

Page 138

1. 7/10
3. 34/100
5. 78/100
7. 5/100
9. 236/1000
11. 5/1000
13. 2464/10000
15. 1 4/10
17. 1 3/1000

Page 139

1. .6
3. .5
5. .005
7. .163
9. .2666
11. .7
13. .007
15. .0014
17. 2.4
19. 2.146

Page 141

1. 1/6
3. 21/100
5. 7 7/8
7. 4.484
9. .00021
11. 340.95096

Page 143

1. .2625
3. 3.4
5. .04375

Page 145

1. 2.76
3. .89
5. 1.00
7. 2.46
9. 7.05

1. 4.2
3. 13.4
5. 3.4
7. .8
9. 3.14

Page 146

1. .176
3. .642
5. .840
7. .002

Page 146 cont

9. 1496.00
11. 12.8

Page 151

1. 10
3. 1000
5. 10
7. 10
9. 100

Page 153

1. 1.3
3. 13.3
5. 8249.
7. 3.47
9. 1.37

Pages 157–159

1. (a) 11 1/5 (b) 11.2
3. (a) 16 1/2 (b) 16.5
5. (a) 30/7 (b) 4.29
7. (a) 57/32 (b) 1.78

Page 160

1. (a) .145
   (b) 1.76
   (c) .00843
   (d) .028

Page 161

1. .36
3. 2.56
5. .375
7. 1.16
9. .2
11. .5
13. .026
1. 25
3. 75
5. 290
7. 41
9. 6.25

Page 161 cont

11. 35
13. 3500

Page 162

1. .5
3. .05
5. .0005
7. .00023
9. .25
11. .00033
13. 25

Pages 163–164

1. 211.12
3. 110.40
5. 196

Page 165

1. 75
3. 85.7

Pages 166–168

1. 6.32
3. 6.5625
5. 96.1
7. 25
9. 480
11. $38.80

Page 170

3. $71.36
5. 50

Chapter 13, Page 177

1. −14
3. 13
5. 8
7. −11

Page 178

1. 10
3. 141
5. 26
7. 326
9. −129

Page 180

1. 8
3. 4
5. −6
7. 0
9. −8
11. −11
13. −472
15. −14

Page 181

1. 5
3. 8
5. 21
7. −3
9. 3
11. 4
13. 42

Page 185

1. 2
3. 16
5. −8
7. 4
9. −20

Page 187

3. 1
5. 10
7. 19
9. 6

Page 190

1. 10
3. −8
5. −2

Page 190 cont

7. 8
9. −19
11. 8
13. 14
15. 9

Pages 193–194

1. −42
3. −30
5. 48
7. −48
9. −72
11. −238
13. 60

Pages 195–197

1. −4
3. 5
5. −10
7. −3
9. −8
11. −4
13. −8
15. −9

Chapter 14, Page 201

2. 27.6 = 27 6/10 = 27 6/10

Page 202

1. 2/5 = .4
3. 6/20 = .3
5. 2/3 = .666 . . . .

Pages 205–206

1. $.\overline{142857}$
3. $5.00\overline{0}$
5. $-.8\overline{0}$
7. $.51\overline{51}$

Page 207

1. 6 1/4
3. 3/5
5. 231/1000

Page 208

1. $1.33\overline{3}$
3. $563.\overline{63}$
5. 1

Pages 211–212

1. 25/33
3. 3/11
5. $\frac{571428}{999999}$
6. $\frac{122}{445}$
9. $\frac{151}{300}$

Page 214

1. 9
3. 361
5. 0
7. 196
9. .0196
11. 169
13. 2 1/4
15. 13/16
17. 121
19. 1/4

Page 217

1. $3 \cdot 2$ or 6
3. $7 \cdot 11$ or 77

Page 218

1. 4
3. 49
5. 36
7. 196

Pages 223–224

1. 7.3
3. 1.732
5. 5.5

Pages 225–227

1. 3.3
3. 54
5. 213.7
7. 8.7
9. 3.6

Page 230

1. $2\sqrt{6}$
3. $5\sqrt{2}$
5. $9\sqrt{3}$
7. $6\sqrt{10}$

Page 231

1. 7.1
3. 16.8
5. 22.0
7. 15.4

Page 235

1. $4\sqrt{5}$ or 8.8
3. $2\sqrt{61}$ or 15.6

Page 236–237

1. 42.3 ft.
3. 9.05 miles
5. 24.1 ft.

Page 240

1. 24
3. 30.672
5. (a) 6
   (b) 20
   (c) 30
   (d) 12
7. Commutative
9. 6

Page 241

1. 5
3. 6
5. 1

Page 242

1. two (a + b), c
3. three 5x, 2a, y
5. two (a + b), (c + d)
7. three 2x, 2y, (−3)
9. two (x − y), 2a
11. two 2A, (x + 6)

Pages 244–245

1. 3
3. −1
5. −1
7. 6
9. 15
11. 2
13. 0
15. 3

Page 247

1. 81
3. 35
5. 24
7. 4224

Pages 249–250

1. 53
3. 51
5. 27
7. −11
9. 11
11. 8

Pages 251–252

1. ab
3. a(b + c)
5. $a + x^4$
7. b − A
9. y^2
11. xy^2
13. $\frac{x}{a + b}$
15. 7 − x
17. yz + x
19. 2(a + 2b)
21. $(a + b)^2$
23. x + 5
25. x + 2
27. 5x + 3y
29. 3(m + n)
31. x + 3y
33. x − (a + b)
35. x(a + b)

Pages 257–258

1. 7(x + y)
3. 15(x + 5y)
5. x(7 + 11y)
7. 2(a + 1)
9. 10x(1 + 10y)
11. 3(5a + 1)
13. $x^2(3y + w)$
15. $13x^2(xy + 13z)$
17. $3x^2(x + y)$
19. 28(w + 7x)

Page 259

1. 2x(1 + 3y + 4a)
3. 7(x + 5a + 2y + 6z)
5. 2(3ab + 2b + 1)
7. (x + y)(7 + 2a)

Page 260

1. (x + y)(2 + a)
3. (5a + 1)(b + 2)
5. (cx + y)(ab + 1)
7. (x + y)(3 + a + 1)
9. (72 + x)(3)

Page 263

1. 7a
3. 9x
5. −5x

Page 263 cont

7. 28 pq
9. 12(a + b)

Chapter 15, Page 264

1. 12x
3. 2x
5. 23ax
7. 7x + 2y
9. 8ay + 3

Pages 265–267

1. 3x + y
3. $16x^2 + 3y$
5. .61x + .37y
7. 3
9. 13ab
11. 2½a + 8b
13. 3abc
15. $4x^3 + 36x$
17. 3a + 2b
19. $-2x^2 + y$
21. 9ax + xy + 4x
23. $5ax^2 + 5y$
25. 4ax + 6by
27. y + 2x
29. 11ax + 4bx + 7ay
31. .62x + 1
33. 0

Pages 268–269

1. identity
3. identity
5. identity
7. identity
9. conditional x = 2; y = 5
11. conditional a = 6
13. conditional y = 0
15. conditional x = 11.3

Page 270

1. yes
3. no
5. yes
7. yes

Pages 273–275

1. x = 15
3. w = 10
5. z = ½
7. x = 2
9. x = 0
11. x = 5
13. w = 17
15. $y = \frac{1}{3}$

Pages 276–277

1. x = 1
3. x = 2
5. x = 7
7. x = 4
9. x = ½
10. x = –5

Pages 278–279

1. identity
3. x = 2
5. x = ½
7. x = 2
9. x = 1

Pages 280–282

1. x = –3
3. x = –1
5. x = 3
7. x = 3
9. x = 1
11. x = –4
13. x = 0

Pages 283–284

1. x = ½
3. x = 2
5. x = 2
7. x = 0
9. x = 5

Pages 285–288

1. $x = -3$
3. $x = -10\frac{2}{3}$
5. $x = 1\frac{1}{2}$
7. $x = -45$
9. $x = -\frac{11}{9}$
11. $x = \frac{7}{2}$
13. $x = -3$
15. $x = 7$
17. $x = -\frac{7}{12}$
19. $x = -\frac{60}{7}$
21. $x = 2$

Page 290

1. $r = 1\frac{1}{2}\%$
2. $I = \$90$
3. $t = 15$
4. $r = 3$

Pages 291–293

1. $t = \frac{I}{Pr}$
3. (a) $l = \frac{P - 2w}{2}$

   (b) $w = \frac{P - 2l}{2}$
5. $m = \frac{fr}{v^2}$
7. $P_2 = \frac{P_1 V_1 T_2}{T_1 V_2}$

INDEX

Metric System

Meter - length (distance) meter = 39.37 inches
gram - wt 454 grams = #1
Litre - vol (liquid) 1 Litre = 1.06 qts

3.77 Litres = 1 gallon
454 grams = 1 pound
908,000 gram = 1 ton
908 Kilograms = 1 ton
1 meter = 1.09 yards
1 meter = 3.28 feet

1.60 Kilometers = 1 mile
0.62 miles = 1 Kilometer
2.54 cm = 1 inch
25.4 mm = 1 inch

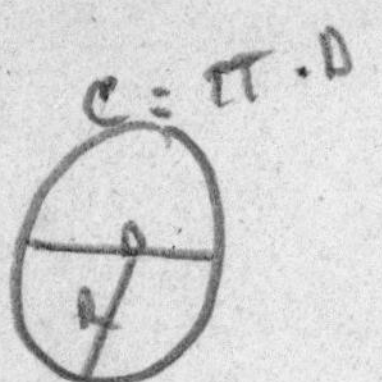

Vol. Box: $L \cdot W \cdot D$ = cu measure

Vol Cylinder $\pi r^2 \cdot h$ = cu. measure

$D = 2 \cdot r$

$A = \pi \cdot r^2$

$C = \pi \cdot D$

$A^2 + b^2 = C^2$